"Melby's book is an engaging demonstration of what holistic anthropology can do. Melby uses multiple complementary levels of analysis, integrating theoretical perspectives from biomedicine, biocultural anthropology, evolutionary biology, and public health, to prompt the reader to look at menopause and aging in a new light."

—**Karen Rosenberg**, *Professor at the Department of Anthropology, University of Delaware, USA*

"Many scholars write about interdisciplinary science, but Melby's accessible book walks the walk! She skillfully weaves insights from her participant-observer research with theory to fully articulate her comprehensive Proximate-Ultimate Biocultural framework. A must-read for all those interested in human development, culture, and their intersections."

—**Julie Hicks Patrick**, *Professor of Life-Span Developmental Psychology at West Virginia University, USA*

REFRAMING AGING

How do we age? Why do we age? How and why does menopause happen? Do different cultures have different approaches and attitudes to, and experiences of, aging and menopause? *Reframing Aging: Insights from Biology and Culture of Midlife Japanese* uses a biocultural framework to try to answer these questions, and gain insights on aging and menopause in Japan, the United States, and beyond.

Drawing on years of fieldwork and lab work in Japan, and over a decade of living and working in Japan at several universities and the National Institute of Health and Nutrition, with follow-up interviews spanning over 20 years, Melissa Melby challenges what are often considered "normal" experiences of aging and menopause. This book introduces a proximate-ultimate biocultural framework to guide the reader through questions of how (proximate) and why (ultimate) we age, and experience menopause, as we do. Drawing insights from evolutionary biology and societal-level phenomena, and the language of lived experience, it explores how cross-cultural variation in expectations, medicalization, collectivism, lifestyles, and other factors may influence how symptoms of aging and menopause are perceived, experienced, and treated.

Reframing Aging: Insights from Biology and Culture of Midlife Japanese offers new approaches and insightful perspectives for students of biological/cultural/medical anthropology, gerontology, Asian studies, women and gender studies, medicine, and public health.

Melissa K. Melby is a medical anthropologist and Professor of Anthropology at the University of Delaware, USA. She has been conducting research on biocultural understandings of, and approaches to, aging and menopause for over 25 years, with a particular focus on Japan.

New Biological Anthropology

Series Editor: Agustín Fuentes, Princeton, USA

Emergent Warfare in Our Evolutionary Past
Nam C. Kim and Marc Kissel

The Promise of Contemporary Primatology
Erin P. Riley

Genetic Ancestry
Our Stories, Our Pasts
Jada Benn Torres and Gabriel A. Torres Colón

The Dialectical Primatologist
The Past, Present and Future of Life in the Hominoid Niche
Nicholas Malone

Reframing Aging
Insights from Biology and Culture of Midlife Japanese
Melissa K. Melby

For more information about this series, please visit: https://www.routledge.com/New-Biological-Anthropology/book-series/NBA

REFRAMING AGING

Insights from Biology and Culture of Midlife Japanese

Melissa K. Melby

Routledge
Taylor & Francis Group
LONDON AND NEW YORK

Designed cover image: serikbaib, Getty Images

First published 2026
by Routledge
4 Park Square, Milton Park, Abingdon, Oxon OX14 4RN

and by Routledge
605 Third Avenue, New York, NY 10158

Routledge is an imprint of the Taylor & Francis Group, an informa business

British Library Cataloguing-in-Publication Data
A catalogue record for this book is available from the British Library

Library of Congress Cataloging-in-Publication Data
Names: Melby, Melissa K. author
Title: Reframing aging: insights from biology and culture of midlife
Japanese / Melissa K. Melby.
Description: New York, NY: Routledge, 2026. |
Series: New biological anthropology | Includes bibliographical references and index.
Identifiers: LCCN 2025008951 (print) | LCCN 2025008952 (ebook) | ISBN
9781032328584 hbk | ISBN 9781041082453 pbk | ISBN 9781003644521 ebk
Subjects: LCSH: Aging—Physiological aspects—Japan | Aging—Psychological
aspects—Japan | Menopause—Physiological aspects—Japan |
Menopause—Psychological aspects—Japan | Middle-aged women—Health and
hygiene—Japan | Middle-aged women—Psychology—Japan | Self-care,
Health—Age factors—Japan
Classification: LCC QP86 .M475 2025 (print) | LCC QP86 (ebook)
LC record available at https://lccn.loc.gov/2025008951
LC ebook record available at https://lccn.loc.gov/2025008952

ISBN: 9781032328584 (hbk)
ISBN: 9781041082453 (pbk)
ISBN: 9781003644521 (ebk)

DOI: 10.4324/9781003644521

Typeset in Sabon
by codeMantra

This book is dedicated to all the participants who allowed me to enter their lives, their homes, and their hearts. May this book do justice to your stories and contributions to understanding aging and menopause.

CONTENTS

FIGURES

TABLES

PREFACE

My interest in the biocultural factors involved in aging and menopause grew out of a long-standing fascination with environmental influences on physiology and health. Many years ago, while studying plant-insect interactions, I came across an article on medicinal plants and decided to study human-plant interactions. My academic journey took me from biology and chemistry, to environment and development, and finally to anthropology through a study of ethnobotany. I then realized that while medicinal plants had the potential to exert great effects on human physiology and health (as evidenced by the numerous powerful drugs that were originally derived from plants), chemicals in food plants might also exhibit significant influence due to the frequencies and quantities in which they are consumed. I searched the literature for examples of cultures with high consumption of food plants with potentially bioactive compounds that might affect health, particularly reproductive functioning.

When I read Margaret Lock's book, *Encounters with Aging*, I was intrigued by her description of the concept of *kōnenki* (Japanese word often translated as menopause, but overlaps considerably with aging), and how it seemed to not quite map to Western biomedical ideas of menopause. She also proposed that Japanese women might have lower rates of menopausal symptoms compared to North Americans because of their culturally influenced *local biology*, influenced by factors such as their diet containing high amounts of soy. Since Japanese women had the longest life expectancy in the world and health profiles that were remarkably better than women in many other countries, I was curious about how their environment (including diet) might play a role. Japan seemed to provide a perfect case study for a biocultural analysis. An examination of the menopausal transition seemed ideal, since women experience a great range of hormones throughout the transition, permitting examination of complex relationships. While significant health concerns about environmental hormones exist, soy phytoestrogens are naturally occurring environmental hormones that appear to have health benefits. And in Japan, a natural experiment in soy phytoestrogen exposure occurs with regional differences in soy consumption, and individual differences within those regions.

I went to Japan intending to use a broad reproductive ecology approach to study *kōnenki*, aging, and the menopausal transition by combining epidemiological, nutritional, neuroendocrine, and cultural methods and data. But along the way, I discovered that people's language might provide even more important clues to interesting biology, and the lines between biology and culture began to blur. By measuring hormone levels, symptomatology, lifestyle, and life history factors in women aged 45–55, I sought to examine women's midlife transition in Japan from: (1) a proximate biological perspective (using hormone measures), (2) a proximate cultural perspective (using symptom checklists, surveys, interviews, and life histories), (3) an ultimate biological evolutionary and ecological perspective (considering how genes from the gut microbiome might interact with environment in the form of diet and lead to different phenotypes), and (4) an ultimate cultural perspective (by comparing my data to Lock's historical findings in the early 1980s on symptoms of *kōnenki* and sociocultural factors such as medicalization).

This research received considerable funding from NSF and NIH, for which I am grateful. Much of the funding covered measurement of hormones, admittedly the focus of my primary hypotheses. And yet, some of the most significant findings of this research came from simply listening, really listening, to people talk about their lives, symptoms, and experiences, and then asking follow-up questions. A great deal of biomedical research funding on aging and menopause is invested in proximate biological interventions (e.g., hormones and drugs). However, many people in the world may not have socioeconomic resources and access to benefit from these. Other people may prefer more cultural interventions (e.g., social support and policies that change roles and improve access to health-promoting social and environmental determinants such as food, education, and decent working conditions). Populations such as the Japanese may suggest alternative paths to navigating later years of life and menopause with better health outcomes. A focus on environmental and lifestyle interventions may hold greater promise for cost-effective accessible preventive health programs that will improve quality of life in later years for more people.

The research described in this book involved collection of a wide range of data: symptom checklists, abridged dietary information, and views on *kōnenki* from women and men in Kanazawa, Japan ($N = 54$) in 1999; a questionnaire including demographic, health and reproductive history, two-week symptom recall, food frequency, and life events, anthropometric measures (e.g., height, weight, skinfolds), and six months of daily symptom checklists, weekly blood samples with matched 24-hour dietary records, and lengthy interviews about *kōnenki* with women in Kyoto ($N = 68$) and Fukushima ($N = 77$) prefectures, from 2001 to 2003. This fieldwork was followed by return visits to participants, while doing analyses of their blood samples in Tokyo during 2003–2005, analysis from 2005–2007, and then periodic interviews for subsequent studies and informal visits in 2008–2011, 2012–2013, 2015, 2016, 2018, and 2024. I also resided in various cities throughout Japan continuously from 2001 to 2011, and 2012 to 2013, and both of my children were born and raised in Japan for the initial years of their lives.

The initial six-month protocol of daily symptom checklists and weekly blood samples was honestly horrendous. I tried to do it myself for a week and am fairly certain I would have struggled to continue for the entire six months. But the rate of compliance was phenomenal, supported by wonderful collaborators and research assistants, and of course primarily by the dedication of the women themselves. I'd also like to think that my genuine interest in learning about their lives and experiences helped, as demonstrated by the lengths I sometimes went to in order to interview them. Throughout this book, I have included a series of vignettes, or fieldnotes, of particularly striking interactions that occurred during my fieldwork and that illustrate concepts or themes. Here I share a couple fieldnotes that illustrate context and approaches used in my research.

Fieldnotes

Early in my fieldwork, I set out one summer morning for an interview at a woman's house on the outskirts of Kyoto in a place called Katsurazaka. Although familiar with Japanese conventions of changing initial sounds when combined with other words such as ka *to* ga, *and* sa *to* za, *I had not realized that Katsurazaka was the same as Katsura-saka (with saka denoting hill), until I cycled across town and reached the foot of what seemed to be more like a mountain than a hill. These were the days before smart phones and GPS, and instead everyone relied on faxes for maps and the local* kōban *(police box) to give directions – perhaps a sign that there was relatively little crime in Japan for police to occupy their time with. On the corners of most major intersections in a dense city like Kyoto, you could find a* kōban. *There was usually a large map posted on the wall containing the names of every household in the area. I would routinely enter and inform the police officer that I was looking for X-san's house, and they would show me where it was and instruct me on how to find it. Although Google maps might be useful, I doubt they have replaced the knowledge of the local* kōban*!*

By the time I reached the woman's house, I was drenched in sweat – due to cycling up the large foothill of the mountain, on an incredibly hot summer day in Kyoto, which is in a valley famous for its humid summers. The participant opened the door, and we exchanged the initial standard greetings.

Hajimemashite *(Literally: it is the first time [we are meeting]). And I am thinking to myself I must be making a great impression standing here drenched in sweat. After saying my name I said,* "Yoroshiku onegai itashimasu" *(a standard first greeting, but one that literally means "Please be favorable to me.") Perhaps she took it literally, while simultaneously taking pity on me.*

Her next utterance (probably never heard before or since) was, "Before we begin the interview, would you like to have a shower?"

A moment of awkward silence followed. Being one of my first interviews, I did not know if it was appropriate to accept the offer. I thought that for my comfort, and probably for hers as well, I should probably take her up on it. I felt so miserable, and I feared that my state might distract my participant, so I accepted the offer – although I had to get re-dressed into my sweaty clothes.

Despite the awkwardness of the moment, or perhaps even because of it, I believe the efforts I made – cycling up the foothill of a mountain for an interview on an incredibly hot humid day in a Kyoto summer – conveyed, more than words ever could, how interested I was in hearing about her life and experiences.

This story is a great example of the seemingly immediate and genuine bond I often experienced in my fieldwork. My efforts to hear women's stories and lives were rewarded with great details about their experiences, and incredible compliance with a time-consuming research protocol. However, some people took longer to warm up to me...

Fieldnotes

She had agreed to be interviewed at her house. When I arrived, I rang the bell, and the woman slid open the outer door and invited me into the genkan *(entrance way). I stepped inside and waited for her to invite me to step up into the house. Usually, guests remove their shoes at the entrance and then step up, sliding their feet into the slippers neatly lined up at the entrance by the host, and then are shown to a room – traditionally with tatami mats. But the invitation never came. Instead, I conducted the entire interview in her doorway, an odd and slightly uncomfortable experience, with all my documents, scale, skinfold calipers, and lancets for collecting blood spots. I was left to wonder about the interior structure of her home.*

Much later, in a subsequent interview, not only did she invite me into her home, but she lavished gifts upon me – several beautiful prints of women wearing the traditional 12-layered kimono called jūnihitoe *(12 layers), which were worn by women of the Imperial Court in the Heian period. Just as these women had multiple layers of dress, I found that many Japanese, especially in Kyoto, had many layers of formality and* tatemae *(public façade), which took months and sometimes years to peel away to begin to get closer to the* honne *(true feelings). I now have one of these prints hanging in my office to remind me of her, and the many beautiful, complex layers of people's experiences existing below the surface.*

Over time in interviews, we slowly peel away layers, as we do in this book to uncover underlying biological and cultural factors shaping the experience of aging and menopause. As we peel away layers of experience and understanding, it can be helpful to have a framework around which to organize the insights we gain. This book uses the initial fieldwork (and many published articles from this research) as a jumping off point, and incorporates insights from years of additional fieldwork on topics ranging from food and nutrition education and policy, maternal child health, alternative medicine, and most recently the microbiome, as well as decades of presentations and teaching to a wide range of audiences, including students, medical professionals, lay people, and some of the participants themselves.

I have also personally experienced some aging during this time, and hopefully gained some wisdom and insight as well. The framework used in this book is one I have developed over years of teaching about the anthropology of health from multiple complementary levels of analysis, spanning from proximate (answering "how" questions) to ultimate (answering "why" questions) and incorporating biological and cultural explanations. While this book focuses on aging and menopause, the framework and complementary levels of analysis can be applied to many topics.

This book is being published at a time when populations globally are aging rapidly, and the pendulum seems to be swinging back to where Western biomedical views about menopause and hormone therapy were about 30 years ago. Increased openness and conversations about women's health and experiences, the COVID-19 pandemic and rise of telemedicine, political-economic-environmental anxiety, and sociocultural pressures to avoid or delay aging, among other factors, may play a role. There is much talk about menopause and aging and promotion of various interventions and treatments, but perhaps not with a very complete picture of the many factors influencing the range of possible experiences during this period of the life course. This book aims to fill these gaps, and simultaneously provide a theoretical framework that can be applied to phenomena besides aging and menopause.

Language can be tricky, and changes over time. Usage also varies between cultures. When I first began my research, I did not want to use the Japanese term for husband, which literally translates as "master" and so I used "partner" as a foreign loan word – and most participants had no idea what I was referring to… so I reluctantly resigned myself to using "master" as I was there to study the culture, not change it. However, I am aware that my research did change some of the participants' behavior at least temporarily. Some women reported that the burden of having to report everything they ate in a 24-hour period led them to eat foods that were simpler to record on those days of the study. And I did refer one woman to a menopause specialist who was concerned about osteoporosis because her periods stopped when she was in her 30s.

But my participants likely changed me far more than I changed any of them, with one participant's family and home becoming my *jikka* (one's parents' home), and many others providing second homes. Many of them shared their perspectives on having children as an important event in their life histories, which likely influenced my own decision to have kids, and thus my own kids should feel indebted to all these women, otherwise they might not be here today.

Just as my kids (and my career) might not be here were it not for many of my participants, this book would definitely not have seen the light of day without the many people who supported me throughout the research and writing. They are too numerous to list exhaustively, but some of them are mentioned in the acknowledgments. And it seems poetic and humorous that I am publishing this just as I am navigating my own menopausal transition. I had (half-jokingly) set a goal of finishing this book before I reached menopause, but my personal biological clock has definitely been ticking. Every time I got my period, I felt like I got an extension on the manuscript. While my expectations of my own menopausal transition certainly have been shaped by the stories and experiences of my participants, my own menopausal experience might have biased me to emphasize certain points and framings more than others. So, I am happy to report that I have met my personal goal and completed this before any bothersome signs of menopause appeared, although perhaps just barely (and we can only know in retrospect, like so many insights about life).

Newark, Delaware
February 2025

ACKNOWLEDGMENTS

Writing this book was a lengthy process. It is the culmination of decades of research and participant observation in Japan. I have taught and presented ideas in this book on multiple continents, with fascinating discussions and feedback. Many people were instrumental in helping me push this book over the finish line, for which I am deeply indebted.

This book owes its inception to Agustin Fuentes for the invitation and vision that I could build on my published papers and include stories from the field to make this material more accessible.

I set myself a goal of finishing this book before I reached menopause myself, so the year I turned 50, I resorted to what some might call drastic measures... I hired an editor (best decision ever) and had one of my most brilliant ideas (as my ability to sit in front of the computer and write for hours was compromised following multiple concussions and the resulting post-concussion syndrome). The pandemic gave me the insight to use zoom and "teach" my book in a course called "Biocultural Approaches to Aging and Menopause" as a series of 10 hours of lectures and discussions at the Institute for Continued Learning at Roosevelt University outside Chicago. Students were lifelong learners from varied backgrounds, and I am indebted to the students (particularly Pat Kennedy, John and Ruth Sellers, Ilene, Carol and Christopher Melby, and Andrea Zietlow) for engaging with the material and me, asking questions, and sharing insights and feedback. Their enthusiastic feedback helped me refine my arguments and explain complicated concepts. Proficiency with zoom and automatic transcripts being one of the gifts of the pandemic, I used the zoom transcripts to generate a new rough draft.

My editor, Pia Kohler, was instrumental in helping me not only "reframe" the book using my proximate-ultimate biocultural framework, but also my process of writing the book (and perhaps my relationship with the book). She heroically merged my earlier draft with the "course" transcripts so that I had a decent draft to work on. In addition to editorial guidance and feedback, she provided significant moral support throughout the many ups and downs of writing this book. In many ways, she served as my book doula, and I don't think either of us could have anticipated how this "book pregnancy" would at times feel like it would never end… active labor was intense and she was there by my proverbial side throughout.

This book may be one of the last books to be written without the assistance of artificial intelligence – whether that is a good thing or not, only time will tell! And the non-artificial intelligence of many humans contributed to the final book. Many readers (from various disciplines and backgrounds) provided extensive feedback and ensured that it was accessible, clear, and engaging. Of course, I take full responsibility for any errors or unclear, jargon-laden passages.

In the early stages of this book, several people from the Reed College Alumni Writer's Workshop (Chris Lydgate, Kristin Coan, Margarett Waterbury, Stephanie Globus, and Lise Funderburg) provided me with feedback, and helped me become more comfortable with sharing early-stage drafts with readers. So subsequent readers have them to thank for the state of the rough drafts I shared!

The following people generously provided feedback on and assistance with:

1 Research articles: Michael Griffin served as my primary editor for almost all the research papers upon which this book draws.
2 Early drafts: Kathy Dettwyler read numerous sections and drafts and provided extensive feedback and stimulating conversations.
3 The entire book: Pia Kohler (numerous times), Julie Hicks Patrick, Carey Booth, Carol and Christopher Melby, Jan Zeserson, Maya Weaver, Alana Hubbard, and Elana Knopp. I would like to give a special mention to my students Maya and Alana for their close reading and detailed comments – working through their comments was like having engaging conversations, and since they represent much of my intended audience of students, their feedback was invaluable.
4 Select chapters: Cecilia Isaacs-Blundin, Susan Pickard, Jerilynn Prior, Karen Rosenberg, Lynnette Leidy Sievert.
5 Japanese translation assistance: Elizabeth and Yasunari Izaki.
6 Artwork: Kristen Melby drew the spandrel. Susan Kokura helped make all the figures look consistent and professional.
7 References: Librarian Alison Wessel helped with permissions and gave reference related advice.
8 Publishing: Meagan Simpson, Jodie Collins, and Alyss Barraza answered numerous questions about the publishing process. Ameer Basha and team for production assistance.

9 Opportunities to present: in addition to providing me with regular accountability and support throughout, Julie Hicks Patrick invited me to present these ideas to WVU AGE-ADAR Scholars, who provided great feedback.
10 Physical and mental space to write: Cecilia Isaacs-Blundin and Carey Booth hosted me for week-long writing retreats.

It takes a village to raise a child, and a book is not unlike a child in many ways. Many people in my extended village have supported this project and me through the many years of research and writing. This book would not exist without the research upon which it is based.

Research Participants

First, and foremost, to the many participants in the studies described in this book, I offer sincere thanks for their generosity, dedication, and sacrifice – giving up to 6 months of their lives to the prying eyes of an anthropologist, opening their homes and their hearts to me. My relationships with them have forever changed my life, in positive ways too numerous to count.

Research Assistants

I am greatly indebted to the many Japanese research assistants without whom this work could not have been completed. They include: S. Ushida, C. Komori, and S. Matsueda for assistance with interviews and fieldwork data collection; K. Takeuchi and K. Maizuru for assistance with data entry; N. Nishi, M. Murashima, J. Kita, J. Tada, M. Ito, M. Shirakawa, T. Negishi, Y. Negishi, S. Koyoshi, R. Goto, R. Shimura, E. Nishitani, and H. Chie for assistance with dietary record data entry and analysis.

Research Collaborators

This research could not have been conducted without the assistance and support of my numerous Japanese collaborators (affiliations at the time of collaboration): Professor H. Shimizu and Dr. C. Nagata (Department of Public Health, Gifu University School of Medicine); Professor and Chairman Y. Watanabe and colleagues (Department of Social Medicine & Cultural Sciences, Research Institute for Neurological Diseases & Geriatrics, Kyoto Prefectural University of Medicine); Dr. K. Tominaga and staff (Loma Linda Clinic, Koriyama, Fukushima); and Professor and Director-General S. Watanabe (Tokyo University of Agriculture, Tokyo, Japan; National Institute of Health and Nutrition, Tokyo, Japan); Professor and Dr. K. Watanabe (Keio University School of Medicine).

For training in laboratory methods, advice on method development, and feedback on analyses, I would like to thank the following people: J. Stallings for training in blood spot methods; K. Ouchi for instruction in HPLC-CEAD; M. Uehara for assistance with TR-FIA and for sharing laboratory equipment and reagents; H. Adlercreutz and A. Samaletdin for training in TR-FIA of phytoestrogens; J. Peñalvo for careful reading, extensive comments, and advice on analysis of HPLC-CEAD method data.

Thank you to S. Pattison-Cisna, M. Hays, and S. Bridges for assistance with grant and project support while I was in the field.

Institutional and Financial Support

The work described in this book was carried out while I was affiliated with and supported by the following institutions: Department of Anthropology, Emory University, Atlanta, GA, USA; Princeton-in-Ishikawa program in Kanazawa, Japan; Department of Social Medicine & Cultural Sciences, Research Institute for Neurological Diseases & Geriatrics, Kyoto Prefectural University of Medicine, Kyoto, Japan; Loma Linda Clinic, Koriyama, Fukushima, Japan; the Laboratory of Public Health Nutrition and Epidemiology, Tokyo University of Agriculture, Japan; National Institute of Health and Nutrition, Tokyo, Japan; University of Delaware, Newark, DE, USA.

Numerous funding sources are gratefully acknowledged and include: Woodrow Wilson – Johnson & Johnson Fund for Women's Health; Wenner-Gren Foundation for Anthropological Research; Emory University Graduate School of Arts and Sciences; U.S. National Science Foundation (NSF) (BCS0117097); U.S. National Institute of Health (NIH) National Center for Complementary and Alternative Medicine (NCCAM) National Research Service Award (NRSA) (1 F31 AT01041-01); Mary Ellen A. Sarbaugh Grant for Research on Women's Health; Japanese Ministry of Health, Labor and Welfare (H13-Health-015, H15-Cancer Prevention-061); Fuji Foundation for Protein Research, Fuji Oil Co. (Osaka, Japan); Japanese Ministry of Education, Sports and Culture, CIFAR co-director funds; University of Delaware support during sabbaticals and research leaves.

Mentors

A great debt of thanks goes to my primary research mentors – in all things related to *kōnenki*: Margaret Lock whose book inspired the research and over the years shared not only her questionnaire and Japanese contacts, but also invitations to write an important literature review and to present at a CIFAR Humans & the Microbiome program meeting, all of which changed the trajectory of my career; and in all things related to soy phytoestrogens, HPLC, Japanese public health, and nutritional epidemiology: Shaw Watanabe, who gave me unrestricted access to numerous laboratory instruments, scientific opportunities, and students to assist with data entry and analysis. Thanks also go to other mentors including: Jan Zeserson, for constructive feedback on my early research and for sharing insights, and then reconnecting many years later and reading the entire book manuscript; Margaret Moloney for introducing me to qualitative interviewing techniques and giving me the opportunity to do menopause research in the United States; Carol Worthman for teaching me about reproductive ecology; Pat Whitten for teaching me about environmental hormones and phytoestrogens; Neal Smith for teaching me about evolutionary psychology and complementary levels of analysis; Mel Konner for mentoring me in explorations of sex differences and critiques of biomedicine; Michelle Lampl for her intellectual rigor and mentoring me in writing about complex biocultural topics; and Peter Brown for teaching me (how to teach and write) about medical anthropology.

Friends and Family

There are many people who provided support, encouragement, and assistance in various forms throughout the many years of this project – it is impossible to mention everyone, but I am deeply grateful to all of you.

In Japan, many people became like family to me and offered second homes as well as showed me great kindness, introduced me to many aspects of Japanese culture, and helped my research: MISHOP members, Kazue and Eiichi Harada and family, Saeko Ushida and family, Keiko Maizuru, Mihoko Umegaki, Masa Omichi, Hanaoka-sensei and family, Hatsue and Yuji Munakata and family including Yumi and Atsuko and their families and Bababa and Jijiji, Kunihiko Tominaga and family, Rika Suzuki and family, Mashiko-san and family, Hayata-sensei, Shaw Watanabe, Michiko Kaihoku, and Naoko Ishiwata.

Outside Japan, various forms of support were provided during this long process by Kim Winter, Eglee Zent, Jim Bailey, Kathy Mote, Ellen Lepine, Heather Healy, Kate Geddie and CIFAR HMB colleagues and friends, among others. Julie Berger Cummings and the community at SITS provided much needed moving meditation and community, and I attribute much of my concussion-recovery to dancing with them. My pathfinders group Katy, Toby, and Annaliese provided regular personal and professional support.

Finally, I want to express my gratitude for my family who have provided never-ending support throughout the many challenges of this process: Michael, Yuuki, and Kenji tolerated my distractions at home and supported me through many years of research and writing; my sister, Kristen, supported my recoveries following significant health challenges on several important occasions; my parents, Carol and Christopher, provided numerous educational and anthropological opportunities that inspired me to do this work, as well as months of surrogate parenting and babysitting to allow me to conduct research and write, and then became students in my class and detailed readers of the manuscript. No words can convey how appreciative I am of everything you have given me.

Permissions

Permission to reprint figures from Melby, M.K. 2005. "Vasomotor symptom prevalence and language of menopause in Japan." *Menopause* 12 (3): 250–257 was granted by Lippincott Williams & Wilkins.

Permission to reprint the figure of the "normal" menstrual cycle from Carr B.R., Wilson J.D. 1987. Disorders of the ovary and female reproductive tract. In: Braunwald E, Isselbacher KJ, Petersdorf RG, et al, eds. in Harrison's Principles of Internal Medicine was granted by McGraw Hill.

Permission for the map of Japan was granted by Map Resources.

1

INTRODUCTION

1.1 Why a Book on Aging and Menopause?

How do we age? Why do we age? How and why does menopause happen? Do different cultures have different approaches and attitudes to, and experiences of, aging and menopause? Are some more relaxed and less anxious, or maybe even excited, about the changes aging and menopause bring? What is the role of diet, and are there other biocultural factors that are important? These are some of the questions I explore in this book on various levels ranging from individual to population. There are many phenomena that biocultural approaches seek to understand, and the term "biocultural" itself exhibits considerable variation in usage and definitions.[1] I use this term here to mean interactions between biology and culture that may change either or both. For example, the interaction between diet and aging is a biocultural phenomenon in that culture influences diet, and diet has biological effects influencing human variation in health and aging, which in turn may shape the culture around diet, bringing it full circle.

Anthropology investigates the causes and effects of human variation across time and space, and medical anthropology focuses on health and illness. Cross-cultural studies have demonstrated that the experience of aging and menopause is not universal nor uniform. Biomedicine is often assumed to be neutral and objective. It is a form of Western-influenced medicine that assumes that illness primarily results from departures from universal biological norms. Anthropologists often examine biomedicine as a professional ethnomedicine dominant in the West and increasingly around the world.

Aging is a biological phenomenon with ancient evolutionary underpinnings, and its expression is strongly influenced by the cultural environment, thereby

DOI: 10.4324/9781003644521-1

lending itself to a biocultural approach that aims to integrate biological and cultural factors and mechanisms in its explanations. Aging is a gradual change in an organism that leads to increased risk of weakness, disease, and death, and thus one that all readers are likely to be concerned with, eventually. But one's interest in, and perceptions about, aging may depend on where one is in the lifelong process. Aging is a diverse process, and people around the world experience it differently. It's often the exceptions that prove the rule. People love to share stories of their old aunts or uncles or grandparents who smoked or drank heavily and yet lived into their 90s, despite the average life expectancy being in the 70s or 80s. We often say they have "good genes" although genes do not appear to play the primary role in deciding how long and healthily someone will live. Additionally, diet, physical activity, social connection, and purpose in life play important roles. In fact, the latter two (social connection and purpose in life) may provide the motivation and opportunities to engage in healthy lifestyle behaviors (such as diet and physical activity). Thus, context is critical. Humans have created a plethora of contexts – biological and cultural – that provide opportunities for studying variation and discerning patterns that may help promote healthier aging and greater quality of life.

Menopause is a phenomenon of reproductive aging, and thus in some ways a subset of aging. Menopause involves the aging of a complex interconnected reproductive system that has connections to, and effects on, many organ systems. But it is also set apart from general somatic (of the body, in contrast to reproductive cells and organs) aging to some extent because reproduction ends decades before the end of life for many women. Reproduction, and its end, is not only a primary focus of evolutionary biology, but a phenomenon with varied cultural meanings, behaviors, and policies associated with it. Menopause is a universal phenomenon that all people with ovaries will experience if they live long enough. I want to be sensitive to the issue of terminology, and clearly define my terms. I use "sex" and the terms male and female to refer to biological categories based on chromosomes (XX vs. XY) and gametes (sperm vs. eggs), and "gender" and the terms women and men to refer to cultural categories that include identity and expression which depend on societal context. But even though sex may seem more clear-cut, chromosomes and certainly genitalia and hormones do not fall into dichotomized categories. Even more so with gender. However, because the people in my primary study all identified as women and had ovaries (based on hormonal analyses), the word "women" in this book will be used to refer to people who are assumed to have ovaries. There is great inter- and intra-population variation not only in physiology and hormones, but also in experiences and attitudes toward menopause. Menopause provokes strong feelings and opinions about what is normal, how it should be managed or treated or accepted, and why it happens at all.

In this book, I examine Japan as a case study to provide a window into biocultural understandings of aging and menopause, exploring how experiences vary, individually and collectively, from the dominant North American biomedical model. Japan leads the world in life expectancy and has culturally influenced "local biologies" that challenge dominant Western biomedical understandings of aging and menopause. I was inspired by Margaret Lock's book, *Encounters with Aging*,[2] to explore biocultural underpinnings of differences in menopausal symptoms and experiences she observed between Japan and North America. In this book, I approach the phenomena of aging and menopause from an anthropological perspective that integrates biological and cultural perspectives and levels of understanding ranging from the how? (proximate) to the why? (ultimate) to attain a more holistic, nuanced understanding of lived experiences of aging and menopause.

My work and interest in Japan stem from my early training in evolutionary biology and chemistry and my interest in ethnobotany and ethnomedicine, first looking at medicinal plants and then dietary plants, and their effects on physiology. I incorporate reproductive ecology approaches, which examine ecological factors such as diet and their influence on reproductive function and experience. My primary study focused on culture in the form of diet, particularly soy phytoestrogens, and its effects on symptoms of menopause and *kōnenki* (Japanese word that is often translated as menopause but overlaps considerably with aging). In this book, I integrate biochemistry and physiology, evolutionary biology and life history theory, reproductive ecology, cross-cultural comparisons, and public health and medical anthropology.

In 1998, I first went to Japan interested in how hormones originating outside the body (exogenous), in contrast to those produced by the body (endogenous) might influence reproductive physiology. I was intrigued by both "bad" ones like DDT, DES, dioxin – and potentially "good" ones like phytoestrogens (naturally occurring plant estrogens). At the time, there was a growing interest in endocrine disruptors (chemicals that might "disrupt" hormonal physiology) in the environment, particularly in Japan.[3] Ethical guidelines prohibit researchers from intentionally exposing people to "bad" chemicals, so the fields of epidemiology and anthropology generally undertake observational studies. Since the Japanese were already ingesting relatively high quantities of potentially "good" endocrine disruptors in the form of soy, I decided to take advantage of the known regional variation. Relative to the national average soy intake, the Tohoku region was characterized by 22.4% higher intake, in contrast to the 12.3% lower intake in the Kansai region,[4] so I chose field sites in those two regions to maximize variation.

Research was conducted in three primary sites in Japan (Figure 1.1). I went to Kanazawa in 1999 to study Japanese, and while there conducted a study comparing symptoms of women and men.[5] In 2001, I began my larger Kyoto-Fukushima Women's Midlife Health Study (KFWMHS). The work described in this book builds on over two years of fieldwork and two years of lab work conducted in Japan for my dissertation,[6] as well as over 10 years

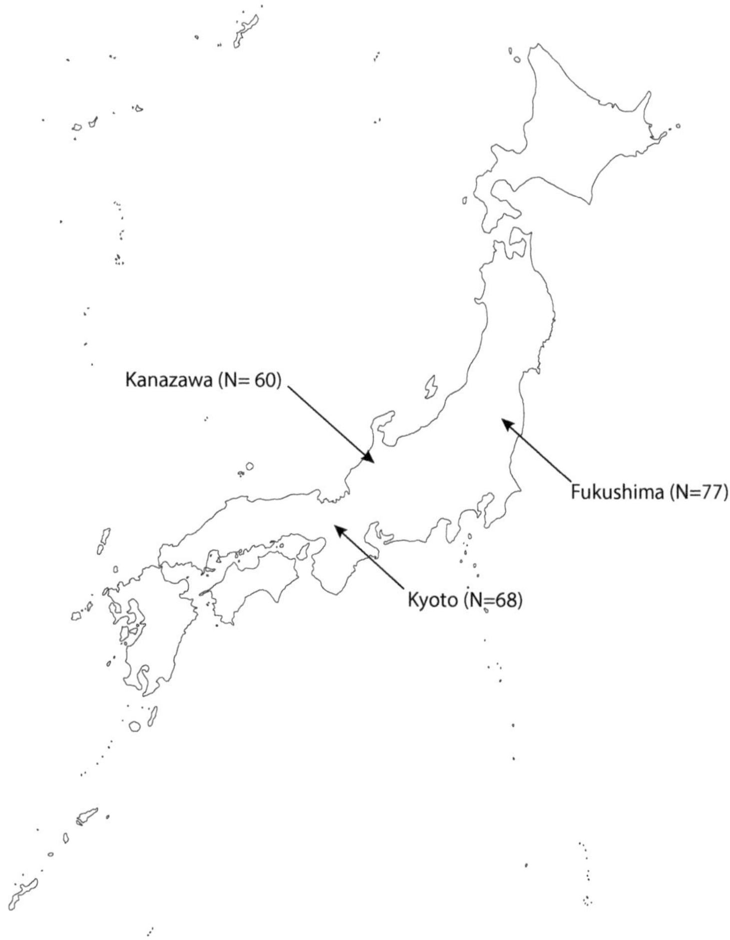

FIGURE 1.1 Map of Japan showing sites where research discussed in this book was conducted. For details about participants and methods (see Melby[5,6]). (reproduced with permission)

living in Japan and researching other topics, including maternal-child health, food and nutrition education and policy, alternative medicine, and microbiome, and over 20 years of studying, teaching, and writing about biocultural anthropology of aging and health. The insights about aging and menopause shared in the book are intended for readers of all ages, particularly those who are interested in aging, women, Japan, or in using biocultural approaches to explore other important biocultural phenomena. Beyond aging and menopause per se, this book uses those topics as a window into applying a biocultural perspective to gain holistic (whole is greater than the sum of its parts, with many components being interconnected) insights into any phenomenon.

This approach can be used to understand topics ranging from obesity to mental health, so I hope that readers will take away tools, including the biocultural framework, that can be applied to many different phenomena of interest and importance.

This book was written to allow people to use it flexibly – like a "choose your own adventure" of sorts. Each part relates to one of the quadrants, and each part is divided into three chapters: an introduction/overview, a chapter on aging, and a chapter on menopause. If you are most interested in: a high-level overview of theoretical issues, then you may focus on the first chapters of each part; aging, then focus on the second chapters of each part; and menopause, the third chapters of each part.

1.2 A Multi-Dimensional Proximate-Ultimate Biocultural Framework

Throughout this book, I deploy a framework I developed that integrates many complementary levels of understanding and approaches. I am an anthropologist by training, but originally specialized as a biochemist. I often begin research and classes with what we know about biology, then move to cultural factors, and finally to integrating the biological and cultural factors.

The framework (as presented in Figure 1.2) provides a structure for understanding different kinds and levels of complementary explanations for a given phenomenon. The *X* axis extends, from left to right, from biological

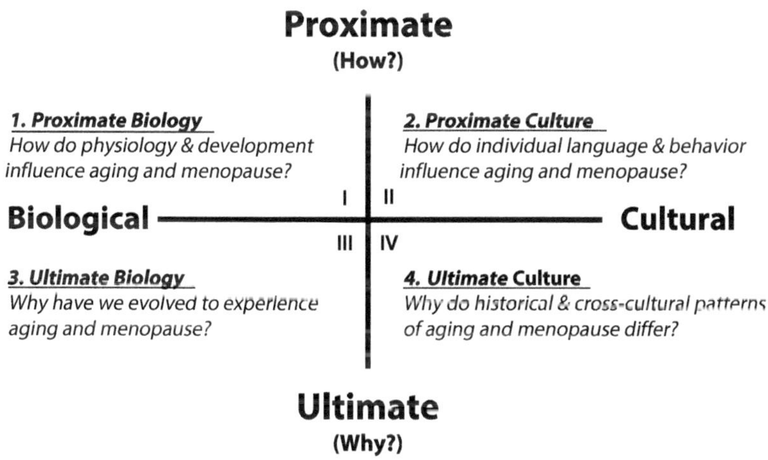

FIGURE 1.2 Proximate-Ultimate Biocultural Framework. Describes the topics covered in each part (corresponding to quadrants of the framework) of the book, examining different levels of explanation from how things work (proximate) to why they exist in the forms they do (ultimate) in domains spanning from biology to culture.

to cultural factors. The *Y* axis extends, from top to bottom, from proximate causes (*how* things work – mechanisms directly underlying behavior or disease) to deeper ultimate causes (*why* things are the way they are – evolutionary or historical origins and the selective processes that have shaped their past and current functions). We can also think of this *Y* axis as moving from an individual to a population level.

The crossing of these axes creates four quadrants. The questions we ask in each can be considered as complementary levels of analysis. Questions that ask how something works, physiologically at the level of the individual, are different from questions that ask why something exists – which can be philosophical or historical, or in the case of biology, evolutionary. This book approaches the issues of aging and menopause, from a holistic, anthropological perspective that integrates biological and cultural perspectives and levels of understanding from the how to the why.

In Part I and Quadrant I (upper left), we consider proximate-biological questions and approaches, and review how something like aging works biologically (physiological mechanisms and development). In Part II and Quadrant II (upper right), we consider proximate-cultural questions, examining how culture works and variations in the expression of aging and menopause (language and experience). In Part III and Quadrant III (lower left), we find ultimate-biological questions that prompt us to ask why something like aging or menopause exists (evolution and adaptation). In Part IV and Quadrant IV (lower right), we turn to ultimate-cultural questions to consider societal and historical reasons for variation in population-level differences in aging and menopause (social determinants and medicalization). Not everything fits neatly into these four different quadrants, but it provides a useful framework for inquiry.

1.3 Biocultural Analysis of Headaches: Example Application of the Framework

Let's consider the example of headaches to help understand how this framework can provide deeper insights and holistic understandings. What are the biological causes of headaches? These could include sinus infection and pressure, muscle tension, fever, and stress. Stressors are biocultural in that they exert biological effects on the body but can be triggered by cultural or psychological factors, demonstrating how the lines between the quadrants may be blurry.

1.3.1 Proximate Biology

When people go to the doctor complaining of headache symptoms, a physician will consider various causes when trying to identify the best course of treatment. On a proximate-biological level, headaches can be caused by chemical activity in your brain, the nerves or blood vessels surrounding your skull, or the muscles of your head and neck (or some combination of these factors).

1.3.2 Proximate Culture

A proximate-cultural approach might consider how headaches are discussed and what variation might signify. Many symptoms, like headaches, are not easily measured physically, so they must be communicated through language. That communication critically influences the diagnosis and treatment. One thing that fascinated me when I first went to Japan is the sheer number of onomatopoeia (words that imitate or suggest sounds they describe). In English, we have onomatopoeia like zigzag or helter skelter. Japanese particularly value descriptive onomatopoeic language such as *giongo* (phenomimes, words mimicking sounds) and *gitaigo* (psychomimes, words depicting states or conditions) to describe symptoms and bodily experiences. If you go to a doctor in Japan for headaches, you are expected to describe it with language that often incorporates onomatopoeia. It is often said that Arctic populations have many words to describe snow because of its importance to their lives. By extension, one might think the Japanese have many headaches, or that they are more important.

Here are some examples of commonly used *gitaigo* for headaches:

1. *piri-piri* = stinging pain; tingling
2. *biri-biri* = feeling an electric shock
3. *jin-jin* = feeling numb; throbbing
4. *gan-gan* = throbbing headache; pounding feeling
5. *chiku-chiku* = prickling; stinging; tingling

Is it possible that these different onomatopoeic words convey different biological experiences associated with different types of headaches? These examples show how difficult it can be to translate concepts and words from different languages.

To native English speakers, such sound words may appear childish, but it is completely appropriate for Japanese adults to describe their symptoms using such words. Similarly, *manga* (graphic novels) are not simply cartoons for children, but can cover "serious" topics such as history or science. Perhaps the structure of the Japanese language (syllabic alphabet and symbolic written characters) renders sound-language and graphic novels more appropriate for use by adults? Could culture, symptoms, and language intersect in ways that provide insights into what is happening with a patient or population?

1.3.3 Ultimate Biology

Moving to a deeper, biological (evolutionary) level, we might ask why headaches exist at all? Can they be adaptive? Why have they evolved? Some people may carry genes that make them more likely to develop headaches, particularly in response to certain kinds of stressors. Some possible ultimate-biological explanations include headache as a: i) defense mechanism, ii) result of conflicts with other organisms, iii) result of novel environmental factors, iv) compromise between genetic harms and benefits, and v) design constraint.

Headaches might be some kind of evolutionary defense mechanism, resulting from conflicts with other organisms such as viruses or bacteria, or they could partly result from a mismatch between our evolved biology and novel environmental factors. One common evolutionary explanation is design constraints: our bodies are constrained by what came before and thus cannot evolve completely novel characteristics if variation does not exist. Evolution is constrained by existing genes, structures, and physiology.

1.3.4 Ultimate Culture

Finally, on an ultimate-cultural level, we could start to look at societal factors such as medicalization, policies, and roles that differ by demographic variables such as gender, ethnicity, and class and their intersections. People's lived experiences can sometimes provide clues to this level, and the following personal story might illustrate some insights to be gained by looking at upstream cultural factors. Similar fieldnotes are included throughout this book, to illustrate, elaborate, and reinforce insights described in the text.

Fieldnotes

Years ago, I began experiencing a very bizarre pain in the back of my head when I was driving, triggered when I looked in the rearview mirror. The headache wasn't related to the moving vehicle or my head moving – it only happened when my eyeballs moved to the right. This could be seen as an evolutionary mismatch with our contemporary environment. Evolutionarily I wouldn't have been looking in a rearview mirror, or doing that kind of eye movement so frequently, so it might never have been triggered. But it impaired my ability to live (and drive safely) in my current environment. Thus, the headache symptom led to an impairment (inability to drive safely). The doctor's challenge was to identify the cause and treat it.

I went to see my primary care physician, an MD PhD and clinic director. I told him I had a weird headache, and with little physical examination, he ordered a CT scan. Thousands of dollars later (paid by insurance fortunately), the scan and tests indicated that the proximate mechanism was not a brain tumor. So, the physician prescribed high doses of the anti-inflammatory naproxen and a tricyclic antidepressant supposedly good for chronic pain (perhaps also hypothesizing there might be a psychological component to my pain, although he never mentioned that). In biomedicine, it is common to try drugs as part of the diagnosis process – if symptoms abate, then we conclude the drug target was the cause. But given multiple drug effects, and possible placebo effects, this may be too simplistic.

After a few weeks on the drugs, I began noticing that I bruised easily. I informed my physician about my concerns, but he said I might have to be on the medicines for the rest of my life. I was unhappy with that suggestion, given that the headaches seemed to come from nowhere. I said I wanted to try some alternatives and asked for a referral to an acupuncturist. His immediate response was "I can't send you to a witch doctor!"

In Georgia, where I was living at the time, the only people that could practice acupuncture legally were medical doctors, making my physician's stereotype even more striking to me. I told my physician that I intended to try acupuncture – which has evidence for helping with chronic pain (for which Western medicine has not often been effective). He refused to give me a referral. I told him that without a referral I would have to pay out of pocket, so he would be causing me additional financial burden. He still refused.

I will confess that out of desperation I resorted to cultural stereotyping and looked in the phone book (yes, that is how long ago this was) to find a chiropractor with an Asian-sounding name, in the hopes that perhaps they would be able to do some acupuncture-like treatment. I found one who informed me that legally he couldn't do acupuncture but could do external electrical stimulation.

After one of the most thorough physical examinations I have ever had, he said "I think I know exactly what's going on with your headaches, and in fact, I think I can replicate it on the other side of your head." I was intrigued and asked him to demonstrate. Using his fingers, he put pressure on a couple of parts of my head and then told me to look to the left in an imaginary rearview mirror. Indeed – it was the exact same pain on the left side of my head that I felt on my right side when looking to the right. How could my primary care physician not have any idea and tell me I was going to have to be on drugs for the rest of my life when this chiropractor so easily replicated the pain on the other side? The chiropractor explained that he thought I had a pinched cranial nerve (the proximate mechanism), and said he could probably fix it in a few sessions. Sure enough, a few sessions, each costing about US$50 in today's dollars, and the pain was gone.

While talking to him over the three sessions (each of which consisted of more time than I ever spent with my physician), discussing what might be causing this cranial nerve to be pinched, he helped me figure out that I was probably clenching my jaw at night.

He asked me if I had any stress in my life (aha – perhaps he was a witch doctor – in the sense of dealing with larger social factors). At the time I was having significant interpersonal stress, which likely led to me clenching my teeth while sleeping, which in turn triggered the pinched cranial nerve and the bizarre headache. So, my "witch doctor" helped me figure out the underlying interpersonal social cause of my headache. I was "cured" after three sessions. I addressed my interpersonal stress, and the headache has never returned.

Returning to our framework, let's examine my personal headache story. On a *proximate-biological level*, headaches are mediated by irritation or stimulation of nerves that connect to the cervical spine structure that receives sensory information from the nerves. Mine was caused by a pinched cranial nerve. But what does that tell us about the "ultimate" cause of our headaches? For example, why we have them at some times but not others, why some people are more prone to them, or why, in my case, the headaches seemed to start overnight and be triggered only in very specific contexts?

On a *proximate-cultural level*, causes such as somatization (expression of emotional distress as bodily symptoms) of stress could be explored, although American culture often is dismissive of psychosomatic illness. Headaches are invisible symptoms, and while they may be caused by interpersonal stress, they can also be used as excuses to avoid interpersonal relations, thereby also potentially causing more interpersonal stress. Headaches as a trope to avoid sex are well-recognized in the commonly uttered phrase "Not tonight honey, I have a headache."

On an *ultimate-biological (evolutionary) level*, perhaps my "selfish genes" and evolved emotional mechanisms may have been pushing me to feel stressed and clench my jaw at night. While my headaches could be treated with medications to interrupt the physiological pain mechanism, long-term health (not to mention minimizing side effects of the medications) might have been better served if deeper, underlying causes, including psychosocial factors that led to me clenching my teeth, might have been explored.

My doctor's stereotype that acupuncturists are "witch doctors" is at the *ultimate-cultural level* of explanation. On this ultimate cultural level, we explore societal factors such as medicalization, policies, and roles. Traditionally in many cultures, people believed in the power of magic to influence behavior and health, and witch doctors were respected people that could help you deal with social factors that might be contributing to your manifestation of illness. Witch doctors often look for spiritual, or psychosocial causes. It is not hard to imagine that a witch doctor, from another time or culture, might identify potential issues in my life such as role conflict or feeling socially disconnected.

Just like my headache, a lot of symptoms and signs of aging are hidden – pain associated with lower back injuries, arthritis or osteoporotic fractures, and early cognitive and neurological impairments. These are often invisible disabilities. In interactions with the medical system, people commonly do not have their experiences and symptoms validated or taken seriously enough by health providers to spend the time to investigate and identify the cause. It may be easier or more efficient to simply prescribe pain medicines, without concern for long-term effects, and never dig deep enough to discern the "ultimate" causes.

1.4 Local Biology and Diet

While the framework shows four distinct quadrants, the interconnections and overlaps are essential to integrating all these perspectives. To understand where inter- as well as intra-population variation comes from, we must examine biological and cultural factors that potentially contribute to "local biologies." Medical anthropologist Margaret Lock used the concept of "local biology" to propose

> not only that cultural beliefs influence the construction, experience, and interpretation of aging and other biological processes but that biological difference – sometimes obvious, at other times very subtle – molds and contains the subjective experience of individuals and the creation of cultural interpretations.
>
> *(p.39)*[2]

Lock argued that the interaction

> between culture and biology implies that we must contextualize interpretations about the body not only as products of local histories, knowledge, and politics but also as local biologies [and] therefore, that we cannot assume the universality of either the menopausal body or the Menopausal Woman.
>
> *(p.39)*[2]

Lock originally used the concept of local biology to challenge the idea of a "universal" biomedical body as the norm. In her book, *Encounters with Aging*, she compared data from midlife women in Japan, Canada, and the United States. She explored how "local" cultural factors might influence individual bodies and result in variation in local biologies. These local biologies lead to different subjective experiences of aging and the menopausal transition. The idea of local biology challenged biomedical concepts of universal biology, which is almost always a standardized white body for which medications are developed and prescribed. But the concept also challenged anthropology by countering the idea that cross-cultural variation is only socially constructed. Some anthropologists reacted to Lock's idea of local biology with anxiety and criticism that it was dangerous and akin to "playing with the fire" of thinking that human behavior and traits are determined by genes and physiology (biological determinism).

However, Lock argues for a need to move *beyond* biological and genetic determinism, *and beyond* cultural determinism, toward a more integrated understanding of how biology and culture interact.[7] In terms of interconnections, epigenetics (how environmental factors turn genes on and off and influence biological variation) may help bridge these two perspectives, by providing clues as to how cultural differences, for example in the form of diet or even psychological stressors, become embodied.[8]

The concept of local biology challenges the nature-nurture divide, and undermines or problematizes the debate, providing a way forward to integrate what are often seen as opposing views. Similarly, the framework presented in this book aims to demonstrate how explanations on multiple levels can be complementary and integrated for a more holistic understanding.

I have long been interested in culture in the form of diet, particularly components such as soy phytoestrogens which are environmental factors that may influence reproductive physiology. What people eat varies tremendously around the world and is influenced by what grows in their environment, as well as cultural evaluations of what constitutes food and what does not. Diet has significant biological influence, not just via macro- and micro-nutrients, but also in terms of the impact on the microbes that live in our gut and their effects on our bodies. Thus, biological differences can be influenced by culture, in the form of diet, and also microbial "cultures." But culture also includes factors such as societal attitudes, especially attitudes toward aging. If a culture values getting older, many of the symptoms of aging might not be experienced as something depressing or hard, but rather as "hard won" by surviving long enough to become a "living treasure." This book explores Japan as a window into proximate and ultimate biocultural explanations of aging and menopause.

References

1. Wiley AS, Cullin JM. 2016. What do anthropologists mean when they use the term biocultural? *Am Anthropol.* 118(3):554–569. https://doi.org/10.1111/aman.12608

2. Lock M. 1993. *Encounters with Aging: Mythologies of Menopause in Japan and North America*. Berkeley, CA: University of California Press.
3. Schug TT, Johnson AF et al. 2016. Minireview: Endocrine disruptors: Past lessons and future directions. *Mol Endocrinol*. 30(8):833–847. https://doi.org/10.1210/me.2016-1096
4. Japanese Ministry of Health. 1996. *Kokumin eiyoo no genjoo - Heisei rokunen kokumin eiyoo choosa seiseki. (The Present State of Nutrition in Japan: The Report on the 1994 National Nutrition Survey)*. Tokyo: Japanese Ministry of Health.
5. Melby MK. 2006. Climacteric symptoms among Japanese women and men: Comparison of four symptom checklists. *Climacteric*. 9:298–304. https://doi.org/10.1080/13697130600868653
6. Melby MK. 2007. *Biocultural Approaches to the Study of the Local Biology of Menopause in Japan* [Ph.D. Dissertation]. United States – Georgia, Emory University.
7. Niewöhner J, Lock M. 2018. Situating local biologies: Anthropological perspectives on environment/human entanglements. *BioSocieties*. 13(4):681–697. https://doi.org/10.1057/s41292-017-0089-5
8. Lock M. 2017. Recovering the body. *Ann Rev Anthropol*. 46:1–14. https://doi.org/10.1146/annurev-anthro-102116-041253

PART I

Proximate Biology

2

PROXIMATE BIOLOGY

Introduction

Questions relating to how things work biologically (e.g., physiological mechanisms and development), as captured in the proximate-biological quadrant (Figure 2.1), often involve immediate concerns such as the signs and symptoms of aging and menopause, and the mechanisms of how they occur. Symptoms are what usually lead people to seek medical advice or healing interventions, and proximate mechanisms are the point of intervention by biomedicine. In Part I, we start by assuming that proximate-biological descriptions are "universal" and discuss the "textbook cases" of aging and

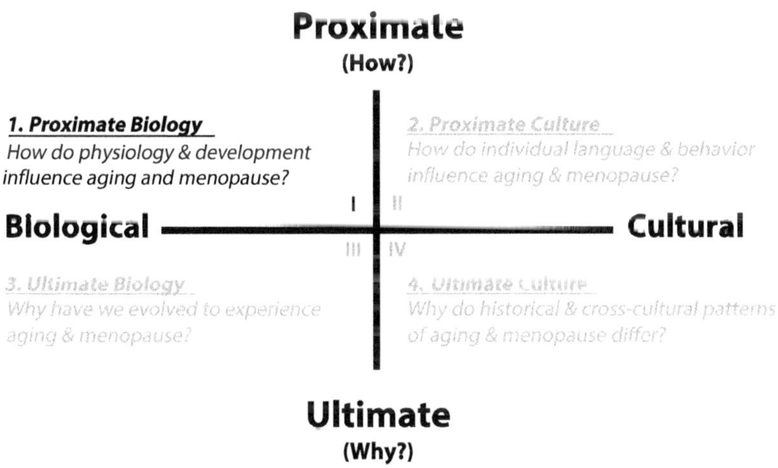

FIGURE 2.1 Proximate Biology Quadrant I of the proximate-ultimate biocultural framework introduced in Chapter 1.

DOI: 10.4324/9781003644521-3

menopause in Chapters 3 and 4. As we move to the other quadrants, we will challenge and problematize that assumption by asking whose biology is considered "textbook" informed by a broader cultural perspective. Spoiler alert: these "textbook" cases are usually WEIRD (acronym for Western, Educated, Industrialized, Rich, Democratic societies), white, and usually male (although in the case of menopause, female).

2.1 Factors Influencing Aging

Perhaps surprisingly, there are no standard or widely accepted definitions of "aging-related" disease. If defined as "diseases that exhibit an exponential association between age and incidence" then "aging-related" diseases include the following: dementia, stroke, and ischemic heart disease.[1] Consensus is lacking on standard biomarkers of aging that should be used in research to assess population differences and effectiveness of interventions.[2] Many hormones change, both in patterns of secretion and in sensitivity to feedback, over the life course and thus may play a role in development and aging.[3] Additionally, the human microbiome (genome of microbes living in or on the human body) also appears to play a role in modulating healthy aging.[4] Furthermore, epidemiological studies have identified behavioral and individual risk factors that are associated with functional decline including: increased age, low physical activity, smoking, extreme (low and high) body mass index (BMI, although this is a flawed measure because it does not distinguish between fat and muscle, and associations with health risks are not as strong as people believe[5]), heavy alcohol consumption, lower socioeconomic status, high medication use, poor self-rated health, and decreased social contacts, in addition to numerous chronic health conditions (lifestyle and larger environmental factors are discussed in Part IV).[6]

Proximate-biological explanations draw from several disciplines including: development, entailing growth and differentiation over the life course; physiology, concerned with functions of living organisms and their parts; and epidemiology, examining the incidence and distribution of health-related outcomes and the biological and cultural factors influencing them.

Many different approaches, ranging from symptom surveys to hormonal biomarkers, are available to study proximate mechanisms of aging. Studies follow individuals longitudinally across the life course or conduct comparative cross-sectional studies of individuals from different age groups, populations, or even species. Animal experiments are often used to test hypotheses, and recent studies that explored the effects of moving gut microbes from young to old animals and vice versa led to the suggestion to replace the phrase "young at heart" (title of 1953 Frank Sinatra song) with "young at gut."[7] Let's see if that phrase catches on!

Approaches are influenced by the focus of study. In medical anthropology, distinctions are made between symptoms, disease, and illness. Symptoms, such as pain, are what bring people to the doctor. Disease is what the doctor, with the help of various diagnostic tests, usually labels. Illness involves social-level phenomena: how people treat a person with symptoms or a disease differently, particularly if they have outward signs of disease such as curvature of the spine. Illness involves role expectations, particularly that of the sick role, or how a sick person with a particular disease is expected to behave. This is particularly true for visible disabilities, but also for invisible disabilities, such as mental illness or chronic conditions.

2.2.1 *Correlation vs. Causation*

Observational studies document associations and correlations but cannot prove causation, although they can hint at it and provide a basis for hypothesis-generation. For example, in human microbiome studies, associations are often found between certain organisms and certain disease states (e.g., obesity, inflammatory bowel disease, diabetes, and Parkinson's). But without

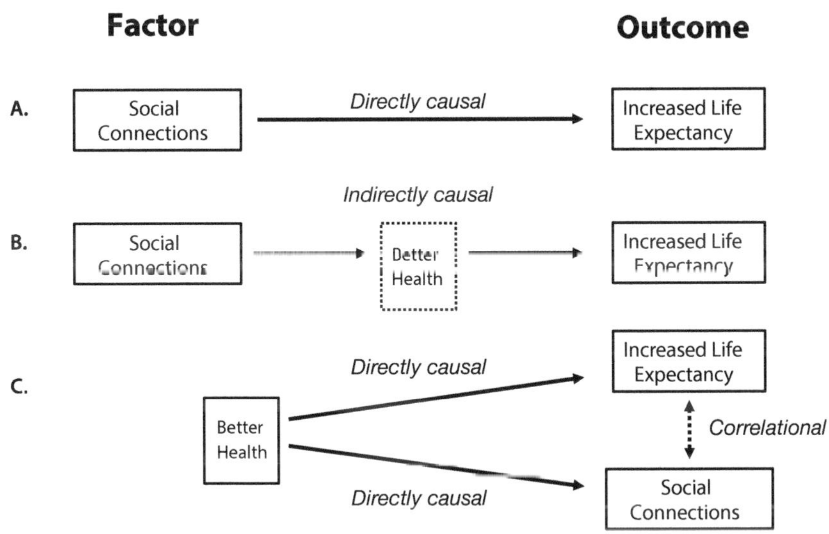

FIGURE 2.2 Relationship between "Social Connections" and "Increased Life Expectancy": Directly vs. indirectly causal vs. correlational relationships. (a) Social connections lead to increased life expectancy. (b) Social connections, through better health, lead to increased life expectancy. (c) Better health leads independently to social connections and increased life expectancy. Social connections and life expectancy are correlated, but not causally related. Modified from Howick et al.[8]

experimental tests (usually in animal models), it is difficult to identify the direction of causality, as microbiome disruption (dysbiosis) may cause disease, and disease may cause microbiome dysbiosis. Similarly, associations between social connections and increased life expectancy have led people to suggest a causal relationship, with social connections "driving" increased life expectancy, but it is important to consider mediating factors and correlations that might confound our understanding of how things work (Figure 2.2).[8]

2.2 Reproductive Ecology

The field of reproductive ecology examines the proximate biology of reproduction and the variation produced from environmental influences, including demographic and ecological factors such as diet and exercise.[9,10] Many studies have focused on "natural fertility" populations, which do not intentionally limit or control fertility, be it through culturally prescribed later marriage or through technological birth control.[11] The outcomes of interest are fecundity (capability of individuals to produce offspring) and fertility (number of offspring produced by individuals). However, "fertility" is often used when authors are really talking about "fecundity," such as in medical "(in)fertility" studies that focus on women with "impaired" or "disordered" reproductive physiology such as "ovulatory dysfunction and anovulation" that impacts fecundity. Reproductive ecology perspectives suggest that what is labeled pejoratively in biomedical literature as "impairments" or "disorders" may actually be an evolved flexible, adaptive response of reproductive functioning to various ecological stressors.[12]

Perhaps the lack of awareness of cross-cultural variation and evolutionary history leads Western biomedicine to see problems in "normal" adaptive human variation. This seems similar to biomedicine labeling humans who are unable to digest milk later in life as having the "problem" of lactose intolerance, when the mammalian norm is characterized by production of lactase enzyme (that breaks down lactose sugar) that decreases with age (i.e., lactase impersistence).[13] Biomedicine often pejoratively characterizes breastfeeding as leading to suppression of menstrual periods (lactational amenorrhea), and age-related patterns of reproductive hormones as causing cycles without ovulation (anovulatory cycles due to luteal "insufficiency") when these are also evolutionarily normal reproductive ecology and developmental patterns. In contrast, a reproductive ecology approach to women's health suggests that differences in the number of lifetime menstrual cycles may explain populational differences in reproductive cancers (discussed in more detail in Part III).[14] Similarly, one might propose that populational differences in menopausal symptoms and experience might also be related to reproductive histories, cumulative hormone exposure, and environmental factors.

From a reproductive ecology perspective, Japanese women look more like natural fertility populations in terms of health (with relatively low rates of reproductive cancers and menopausal symptoms), yet their current reproductive patterns (i.e., age at first marriage, age at first birth, parity, and breastfeeding duration) resemble post-industrialized, developed countries such as the United States and those in Western Europe. Lower rates of breast cancers in Japan may hint at some underlying differences in proximate biology, influenced by cultural environmental factors such as dietary soy.

Fieldnotes

When I began my fieldwork, I hypothesized that soy phytoestrogens might influence the experience of menopause, but also wondered how they might influence breast cancer rates. A menopause clinic physician I met with said she thought breast cancer rates were lower in Japan than in the United States because Japanese women had smaller breasts. I had not heard that theory before! While there is little evidence that breast volume correlates with breast cancer risk, breast density does, likely because it impairs mammogram detection (recall correlation vs. causation). There is also evidence that soy exposure during puberty may affect breast cell differentiation and that, in turn, may decrease breast cancer risk. There have been contradictory results about soy and breast cancer in both animal and human studies, and multiple factors may play a role.[15]

Menstrual periods are an external sign of ovarian function, and generally cease around an average age of 50–52 in white women from industrialized countries, with variation by race/ethnicity, demographic, and lifestyle factors.[16] But as any woman who has done intense exercise or been under nutritional stress knows, periods can stop under stress. And as many women who have struggled to get pregnant know, menstrual periods do not always indicate that ovulation or fully functioning cycles have occurred, and the ability to become pregnant or sustain a pregnancy may nevertheless be compromised.

In other words, periods are external signs of the functioning of an extremely complicated physiological system. However, with increasing ecological, behavioral, or constitutional stress, ovarian function becomes increasingly compromised, eventually resulting in the cessation of menstrual periods.[12,17] And these stressors hint at the need to better understand proximate culture (Part II), for example in the form of diet, that may influence proximate biology.

References

1. Le Couteur DG, Thillainadesan J. 2022. What is an aging-related disease? An epidemiological perspective. *J Gerontol A Biol Sci Med Sci.* 77(11):2168–2174. https://doi.org/10.1093/gerona/glac039

2. Moqri M, Herzog C et al. 2023. Biomarkers of aging for the identification and evaluation of longevity interventions. *Cell.* 186(18):3758–3775. https://doi.org/10.1016/j.cell.2023.08.003
3. van den Beld AW, Kaufman JM et al. 2018. The physiology of endocrine systems with ageing. *Lancet Diabetes Endocrinol.* 6(8):647–658. https://doi.org/10.1016/s2213-8587(18)30026-3
4. Ghosh TS, Shanahan F, O'Toole PW. 2022. The gut microbiome as a modulator of healthy ageing. *Nat Rev Gastroenterol Hepatol.* 19(9):565–584. https://doi.org/10.1038/s41575-022-00605-x
5. Sweatt K, Garvey WT, Martins C. 2024. Strengths and limitations of BMI in the diagnosis of obesity: What is the path forward? *Current Obesity Reports.* 13(3):584–595. https://doi.org/10.1007/s13679-024-00580-1
6. Ferrucci L, Giallauria F, Guralnik JM. 2008. Epidemiology of aging. *Radiol Clin North Am.* 46(4):643–652. https://doi.org/10.1016/j.rcl.2008.07.005
7. Narasimhan H, Ren CC et al. 2021. Young at gut-turning back the clock with the gut microbiome. *Microorganisms.* 9(3):555. https://doi.org/10.3390/microorganisms9030555
8. Howick J, Kelly P, Kelly M. 2019. Establishing a causal link between social relationships and health using the Bradford Hill Guidelines. *SSM – Population Health.* 8:100402. https://doi.org/10.1016/j.ssmph.2019.100402
9. Campbell KL, Wood JW. 1994. Introduction: What is human reproductive ecology, and why should we care about studying it? *Ann NY Acad Sci.* 709(1):1–8. https://doi.org/10.1111/j.1749-6632.1994.tb30384.x
10. Ellison PT. 1994. Advances in human reproductive ecology. *Annu Rev Anthropol.* 23:255–275. https://doi.org/10.1146/annurev.an.23.100194.001351
11. Wood JW. 1989. Fecundity and natural fertility in humans. *Oxf Rev Reprod Biol.* 11:61–109.
12. Prior JC. 2022. Adaptive, reversible, hypothalamic reproductive suppression: More than functional hypothalamic amenorrhea. *Front Endocrinol (Lausanne).* 13:893889. https://doi.org/10.3389/fendo.2022.893889
13. Wiley AS. 2016. *Re-Imagining Milk.* New York: Routledge.
14. Eaton SB, Pike MC et al. 1994. Women's reproductive cancers in evolutionary context. *Q Rev Biol.* 69(3):353–367. https://doi.org/10.1086/418650
15. Bondesson M, Gustafsson J-A. 2010. Does consuming isoflavones reduce or increase breast cancer risk? *Genome Med.* 2(12):90. https://doi.org/10.1186/gm211
16. Gold EB. 2011. The timing of the age at which natural menopause occurs. *Obstet Gynecol Clin North Am.* 38(3):425–440. https://doi.org/10.1016/j.ogc.2011.05.002
17. Ellison PT, Panter-Brick C et al. 1993. The ecological context of human ovarian function. *Hum Reprod.* 8(12):2248–2258. https://doi.org/10.1093/oxfordjournals.humrep.a138015

3

PROXIMATE BIOLOGY OF AGING

You don't stop laughing when you grow old, you grow old when you stop laughing.

—George Bernard Shaw

This quote questions how much aging is in the mind or the body and highlights the issue of chronological vs. biological age. Many people I've interviewed on aging complain about various ailments such as bad knees and backs, but at the same time, they're also glad that they have lived long enough to have those ailments. When asked what aging feels like, one person in their 80s said "everything takes longer and is harder and hurts more...but it's better than the alternative," which is death.

3.1 Definitions

Aging is a gradual change in an organism that correlates with an increased risk of weakness, disease, and death. In gerontology and anti-aging medicine, a distinction is made between biological and chronological age. Chronological age measures the amount of time a person has existed on the planet since their birth. Biological age refers to function measured on the level of DNA by mutations, epigenetic changes, and telomere shortening. Biological age is also assessed on the level of cells, tissues, and organs by the presence or absence of symptoms or diseases associated with the physiological decline that happens with age (senescence). Thus, two people who have identical chronological ages of 80 may have very different biological ages due to different rates of

DOI: 10.4324/9781003644521-4

senescence. And both chronological and biological age are quantitative metrics, but what about more qualitative aspects of aging – how old one feels or acts?

The saying that "you're only as old as you feel" has even been put to a legal test in the Netherlands. Emile Ratelband, a Dutch 69-year old, requested the right to change his birth date to 20 years after the date he was born, because he claimed he had the body of a 50-year-old man, and identified as being younger than he currently is.[1] However, the Dutch court, perhaps not surprisingly, rejected Ratelband's appeal.

Life expectancy is defined for a specific population, as the average remaining years of life expected at a given age such as birth, age 20, or age 50 (see Part III for further discussion). Our species' theoretical maximum lifespan is about 125 years.[2] The oldest person recorded was French woman Jeanne Louise Calmet, who died at 122 years 164 days old, although most of the oldest living humans are Japanese.[3] Research is being conducted on genetics, epigenetics, and the microbiome, investigating if it might be possible to extend lifespan by modifying factors that contribute to biological aging. We also need to consider environmental factors (e.g., diet, lifestyle changes, pollution, toxins, endocrine disruptors) that may be contributing to the decreased average number of years a person can expect to live in "full health" not impacted by disability (healthy life expectancy). Whether the planet (or economies and societies) could handle many people staying alive much longer is another question to consider.

3.2 Signs and Symptoms

While there are many images in the media that instill fear of aging (e.g., bed-ridden elderly in nursing homes), there are also news stories about elderly people remaining healthy and active and achieving physical and mental success that many who are younger cannot achieve, even including reproduction (e.g., men in their 90s, women in their 70s with the assistance of reproductive technologies like in-vitro fertilization)! Many cultures also celebrate and revere older members of their communities. While the United States does have a National Senior Citizen's Day (August 21), Japan's day celebrating elderly citizens (Respect for the Aged Day) is a national public holiday, lending much more weight and honor to the day.

As ubiquitous media tell us, some outer signs of aging include: facial wrinkling, age spots on skin (caused mostly by sun exposure when the top surface layer of skin expands with more pigment and develops what looks like a large freckle), thinning hair, and graying hair (generally happens in one's 30s–40s, unless you have genes like mine and then your first gray hair appears at age 13). While graying of hair has not reliably been associated with poor health, not all outer signs of aging are superficial: male-patterned baldness has been associated with greater risk of enlarged prostate[4] and coronary heart disease[5] (though not likely *caused* by loss of hair – just correlated with it; i.e., they share similar underlying factors).

Fieldnotes

Outward signs of aging vary between populations and individuals, due to genes and environmental factors. Japanese women cover up and protect their skin from sun damage by wearing wide-brimmed hats and gloves that cover the entire arm when driving. There are even gloves that attach to bikes so one doesn't have to worry about forgetting them. Some of my older Japanese women friends told me that they have had age spots removed, but my impression is that due to a lifetime of care and protection from sun damage, many older Japanese women have skin that appears almost flawless as they age, like the white-painted porcelain-like face of a geisha. In contrast, many Caucasian-Americans value having tans in the summer and their rate of skin aging may reflect that preference. Thus, culture can influence biology, and this will be explored further in Part II.

Social expectations about aging (discussed further in Part IV) also influence how people treat older people based on signs of aging. For example, in the United States, my gray hair has led to several interesting encounters: a young man asking me if the child I was breastfeeding discreetly (in my 30s) was my grandchild; a woman asking me about the secret to my great skin (probably because she assumed I was much older); and a pharmacist asking me for my Medicare card when I got my COVID booster even though I wouldn't be eligible for more than 15 years! I suspect that such age assessment errors partly result from the fact that most US women with gray hair in their 30s and 40s dye their hair to cover up their gray (until 2020 pandemic lockdowns occurred). Thus, gray hair in women may often serve as a reliable cue for elderhood, reflecting the age when many women finally stop dying their hair, even if that does not accurately reflect biological variation. Of course, particularly in cultures where elders are respected, people may react differently to signs of aging.

Less visible signs of aging, with more serious health implications, include cataracts, hearing impairments, coronary artery disease, and joint pain caused by arthritis. As one person told me, "You know you are aging when you no longer can stand up and sit down without making noise --- no popping and creaking."

3.2.1 *Applying the Framework: Osteoporosis*

Osteoporosis provides an interesting case to examine using a biocultural approach. From a proximate biological perspective, osteoporosis is a disease marked by thinning of bone tissue and loss of bone density, which increases the risk of bone breakage. This has great impacts on quality of life, as people have trouble maintaining physical activity and health, particularly after breaking a hip.

Over time, decreased bone density in the spinal vertebrae can lead to compression fractures, which may occur during routine activities such as bending, coughing, or lifting heavy objects. Accumulation of compression fractures eventually leads to height loss and kyphosis, also sometimes referred to as dowager's hump. This name reflects its association with older females, perhaps hinting at one factor (biological sex) that influences its occurrence. However, men can also experience osteoporosis, and the stereotype of only women having osteoporosis may hinder diagnosis in men.

In Japan, especially in rural areas, it's not uncommon to see women who are severely hunched over (likely from osteoporosis) working in the fields or rice paddies. The fact that they are nevertheless working in the fields suggests that they may have less compromised quality of life compared to many older individuals in the United States with significant mobility issues from osteoporosis and other causes. While the hunched over individuals in Japan likely have pain, they seem to be able to live independently much longer than many older people in the United States.

While the outer signs of osteoporosis often don't manifest until the latter half century of a person's life, even young readers in their late teens and early 20s should be aware of the lifelong processes leading to osteoporosis. Peak bone density is attained around 22 years for females and 25 years for males[6] and influenced by many factors.[7]

In the case of osteoporosis, the *disease* is the physiological changes in bone structure, only confirmed with dual x-ray absorptiometry or other scans. A person may be unaware of a disease state because they have no symptoms, or because symptoms such as back pain can derive from other causes. The *illness* of osteoporosis may be experienced when bone density scans determine osteoporosis, and calcium supplements and weight-bearing exercise are recommended, or drugs such as bisphosphonates are taken. When *symptoms* become more pronounced, with a hunched back or fractures, then the *illness* experience is often characterized by reduced mobility and social interactions and decline in many areas of functioning.

3.3 Mechanisms of Aging

There are many proximate biological (physiological and developmental) theories proposed for how aging occurs.[8] They are not mutually exclusive, because multiple mechanisms can be at work. We focus on several here (although by no means an exhaustive list): (1) wear and tear (degeneration due to damage or error); (2) genes (telomere shortening, somatic mutations, epigenetics, microbial genes); (3) hormones (neuroendocrine mechanisms); and (4) free radicals (oxidative damage). It might be tempting to categorize the above mechanisms as internal vs. external (e.g., genes and hormones vs. lifestyle factors and exposures). Unfortunately, such neat categories are challenging and do not capture the full range of variation in these phenomena. For example, free radicals can come from outside the body but are also generated by endogenous processes. Hormones can be produced inside (endogenous) and outside (exogenous) the body. Wear and tear could include (internal) DNA telomere shortening and mutations that happen over time, as well as free radicals from within and outside the body. Although there is an overlap between these mechanisms, let's review these separately to better understand the range of explanations for aging that have been proposed.

3.3.1 *Wear and Tear*

Cross-cultural comparisons often reveal different apparent rates of aging, but cross-species comparisons highlight different abilities to repair and "slow down" the "wear and tear" associated with aging. Species vary in their regenerative and repair capacities in different bodily systems. Not all organs or systems wear out at the same rate. In humans, our adult teeth have no regenerative capacity, but shark and beaver teeth do. Human leg joints have some regenerative capacity, but salamanders can regrow tails (including spinal cord and nerves!). In humans, the immune system declines with age and cannot be repaired (although microbiome transplants might help someday). Human eyesight declines gradually until it seems to cross a threshold when most of us require reading glasses for presbyopia (inability to focus on near objects). Human brain cells are not easily replaced if damaged, but some plasticity (ability to change phenotype in response to environment) remains in terms of connections.

Most organ systems exhibit gradual aging, but one system exhibits a discrete change in function (at least based on symptoms we can observe). Which one? The reproductive system in human females! More on that in Chapter 4.

3.3.2 *Genetic Mechanisms of Aging*

Genetics is estimated to contribute to 25% of the variation in aging, with the rest coming from lifestyle factors.[9] Genetics can be divided into host genetics, epigenetics, and microbial genetics. Despite many "safety" measures in place to ensure accurate DNA replication, as cells divide over a lifetime, they sometimes acquire random copying mistakes leading to DNA changes (mutations). Most mutations are unimportant and have few noticeable effects, but over time, they can accumulate and may contribute to aging. Similarly, the end of chromosomes (telomeres) shorten with each division, and this shortening correlates with aging. However, such shortening can also be repaired in some cells and some organisms by the enzyme telomerase.

Mutations can also be caused by environmental factors such as radiation and free radicals. While many mutation-causing factors are external to the body (smoking, radiation), some come from within. Mitochondria (believed to be derived from bacteria that invaded our distant ancestors' cells and formed endosymbionts) are considered the powerhouses of the cell and produce much of the energy required by our cells. In the process of energy generation, however, they also create free radicals (see below), which can damage DNA and proteins in the cell, and are associated with aging. So, the ancient bacteria that give us our daily energy, may also end our life! While our bodies have mechanisms that can repair our DNA, these also seem to decline with age. Impaired DNA repair abilities are linked to cancer, a disease commonly associated with age.

In the late 1990s, there was hope that human health problems would be solved once the human genome project was complete. The rationale was that if we knew all the genes involved in causing or increasing predispositions to disease, we could fix them. Did that happen? No. Why not? As many people learn in high school biology, observable traits of organisms (phenotypes) are not solely a function of the genetic makeup of organisms (genotypes), but rather are products of the interaction between genes and the environment, as expressed by the equation:

Phenotype = Genotype + Environment

But what does it really mean for genes and the environment to interact? Some of the mechanistic pathways by which environmental factors can directly or indirectly modify genetic material include epigenetic modifications of DNA, influences from our resident microbiome (through production of metabolites that modulate and influence host gene expression), hormones which can modify gene transcription, and DNA damage by free radicals.

It is not just what is "*in* our DNA" but what is "*on/above* (epi) our DNA" that influences aging. Genes influence an organism when they are "expressed" to instruct cells to create proteins. But gene expression can be turned on or off by molecular switches (epigenetics), and these may play a significant role in aging.[10] For example, identical twins with identical genes do not always age at the same rate and may have different health conditions due to environmental exposures that modify (turn on/off) their genes. Similarly, all cells in your body have a full set of DNA, but different genes are turned on or off in different cells during key developmental windows to make nerve or heart or muscle cells. The proximate biological mechanisms for these epigenetic modifications include: DNA methylation, histone modification, chromatin remodeling, and RNA modification, all of which are involved in regulating the aging process, and hence contribute to aging-related diseases and epigenetically determined lifespan.[11] These epigenetic changes are influenced by environmental factors including toxins and diet. Thus, epigenetic biomarkers may provide more accurate indicators of biological age,[12] similar to how tree rings reflect environmental stressors.

Aging is affected not only by our host genome but also our symbiotic microbial genome (microbiome) that mediates environmental signals. Changes in our microbiome over the life course are associated with age-related disease risk.[13] In animal studies, transplanting microbiomes from old animals to young animals appears to "age" them, and transplanting microbiomes from young to old animals seems to reverse some signs of aging.[14] Thus, the microbiome provides yet another partly genetic path by which environmental exposures may influence physiological processes of aging.

3.3.3 *Hormonal Mechanisms of Aging*

Hormones are produced from various endocrine glands throughout the body. The hypothalamus regulates the pituitary gland, which in turn regulates key hormonal glands (e.g., ovaries, testes, thyroid, and adrenal glands) in complex feedback loops, which often become dysregulated or less functional with increasing age. Estrogen is a hormone that plays an important role in regulating gene transcription (of DNA to RNA) by binding to receptors inside cell nuclei and to estrogen response elements in promoters of genes, and through epigenetic mechanisms, with multiple effects throughout the body. There are many forms of estrogen, including forms produced by the body (endogenous), such as estradiol, estrone, and estriol, and those produced outside of the body (exogenous), such as phytoestrogens (plant estrogens) and xenoestrogens (foreign estrogens).

Not only form, but also timing of hormones is critical in influencing effects. For example, the insulin/IGF-1 hormonal pathway is a complex network that plays a critical role in cell growth and thus early life survival but may take a toll on organisms later in life. This suggests a potentially important role in aging and provides an example of how some genes may have beneficial effects early in life but detrimental effects later in life (antagonistic pleiotropy) (see Part III on Ultimate Biology).[15]

3.3.4 *Free Radicals: Oxidative Damage*

Free radicals are highly chemically reactive molecules that play a central role in aging by directly causing DNA damage, cross-linking of proteins, and contributing to numerous age-related diseases (possibly through epigenetic mechanisms). Free radicals can be created through normal metabolic processes in mitochondria in our cells, through inflammation caused by infection or even stress, and through environmental sources such as air and water pollution, alcohol, tobacco smoke, pesticides, UV light, and radiation.

A free radical is an atom or molecule that has one or more unpaired electrons, making them highly unstable (because paired electrons create stability). They roam the body searching (not consciously) for another electron from another molecule, and in the process create another free radical. This cascade of free radical creation leads to oxidative stress, which damages cells and DNA, contributing to damage characteristic of aging throughout the body, potentially impacting almost every organ system. Because up to 90% of free radicals are created by mitochondria in the process of energy production, some exposure to free radicals is unavoidable with increasing age, but there are ways of preventing or minimizing the damage. Much focus has been on antioxidants (see below), which stop damage by donating an electron to make paired electrons, thereby stopping the chain reaction.

3.4 Treatment and Prevention of Aging

Much of anti-aging medicine has been focused on reversing or slowing signs and symptoms of aging, such as skin aging. Hormonal interventions (e.g., sex hormones) are sometimes marketed as the key to maintaining youth and health.

Animal studies have shown caloric restriction (i.e., reducing caloric intake by 30–40% of normal "healthy" intake) without malnutrition to be an effective anti-aging intervention, increasing average and maximum lifespans in organisms ranging from fruit flies and worms to rodents and primates, with implications for human aging.[16] While mechanisms remain unclear, potential pathways involve changes in energy consumption, decreased free radicals from metabolism, stress resistance, and the insulin/IGF-1 neuroendocrine pathway.

Although this approach has increased life expectancy in non-human animals, caloric restriction may have negative impacts on other aspects of health. For example, dietary interventions like caloric restriction may impair immune function, reproductive health, and fertility. Caloric restriction (in the form of dieting) in children and adolescents can lead to nutritional deficiencies, growth deceleration, and other health risks such as increased risk of osteoporosis.[17] From a life history perspective (see Part III), tradeoffs between reproduction and growth vs. repair of body tissues are real and hard to overcome. Caloric restriction extends life but impairs fertility (see below), suggesting tradeoffs between reproduction and maintenance of the body. For anti-aging medicine, in a world where many people are choosing to have fewer children, exchanging reproductive potential for longer life may seem worthwhile. But these tradeoffs raise questions about the connections between somatic aging (of non-reproductive cells and tissues) and reproductive aging, and caution that other physiological functions may be impaired as well.

Furthermore, observational studies that detect an association with caloric restriction and longevity may not be able to assess confounding variables and may also have inherent biases and risks.[18] Rather than recommend significant caloric restriction, once we better understand the mechanisms by which caloric restriction leads to life expectancy increases, we may be able to intervene in other ways. For example, if caloric restriction acts primarily through reduced creation of free radicals, then antioxidants might be effective. Antioxidants are cornerstones of anti-aging medicine and marketing. However, experiments designed to reverse effects of oxidative damage by giving antioxidants have been inconclusive. This suggests that antioxidants may lend themselves better to prevention than to treatment, as once the damage is done it may be too late. Consumption of colorful fruits and vegetables containing antioxidants such as vitamins C and E and carotenoids are associated with longevity and may help protect healthy cells from damage caused by free radicals. Generally, the more colorful a fruit or vegetable, the higher the antioxidants. Dietary antioxidants may influence DNA methylation (epigenetics) in ways that are beneficial for prevention of chronic disease,[19] but antioxidant supplements may also increase cancer risk. Thus, the proverbial "fountain of youth" has not yet been found and is unlikely to be available in a health food store near you.

3.5 Osteoporosis: How Biology Works

Proximate biological explanations for osteoporosis include anatomical changes and biochemical changes that involve bone creation and breakdown (remodeling), and epidemiological factors associated with bone density and fracture risk. Sex differences exist, with women having higher rates of osteoporosis than men, so sex hormones (such as estrogen) and behavioral and environmental exposures may also play a role.

Bone loss is believed to be caused by lifestyle factors such as low: dietary calcium, exposure to sunlight and vitamin D (which helps absorb and retain calcium), and frequency of weight-bearing exercise. Vitamin K2 and the microbiome may also play a role. If those are insufficient, treatment with bisphosphonate drugs, which slow bone breakdown, have shown success. But both buildup and breakdown of bone are required for healthy bone, so over time bisphosphonates appear to lead to weaker bone (compromised structure) because they decrease breakdown of old bone. Since osteoporosis is more prevalent in women, and risk increases after menopause when estrogen decreases, this correlation suggests (but does not prove) that estrogen should help prevent osteoporosis (see discussion of WHI study in Chapter 4). This correlation often provides incentive for women to consider taking hormone therapy (HT) at menopause.

In the United States, marketing dairy as the primary source of calcium has been very successful,[20] so many people believe milk is essential to preventing osteoporosis. But populations vary in their dairy product intake. Does this mean that populations vary in their rates of osteoporosis? They do. Do you think populations with high intake of dairy products would have higher or lower rates of osteoporosis? The Dutch drink a lot of milk and eat a lot of cheese compared to people in Japan. Who do you think is likely to have higher rates of osteoporosis, the Dutch or Japanese? The answer is very counterintuitive. Within populations, such as the United States, that are over-represented in medical studies, increased calcium and dairy intake is associated with decreased risk of osteoporosis, but cross-culturally, this relationship does not seem to hold. The highest fracture rates, which are a measure of osteoporosis, surprisingly occur in countries with the highest calcium and dairy intakes. And the lowest rates occur in countries with low intakes. This has been referred to as the *calcium paradox*.[21]

Mammals generally stop producing the enzyme lactase (that breaks down the lactose sugar in milk) after weaning, since they no longer need it. But some humans have evolved mutations that allow production of lactase to continue into adulthood, contributing to population and individual variation in the ability to digest milk. Humans with lactose intolerance (or lactase enzyme impersistence) cannot digest the sugar in milk after their bodies stop producing the lactase enzyme that breaks down lactose. They may have considerable discomfort or symptoms when they consume dairy products. For them, obtaining calcium from non-dairy sources is critical. Lactose intolerance is higher among many African-Americans (depending on their ancestry) compared to whites descended from Northern Europe. African-Americans also generally have lower concentrations of vitamin D, but yet also have lower rates of osteoporosis compared to whites. This is called the *osteoporosis paradox*. In Part III, we will explore evolutionary insights into this paradox.

Such population differences tempt people to conclude that race may play a role. But lifestyle factors correlated with race, such as smoking, diet including animal protein intake, and exercise may play more causal roles. Race is a category often used in medical and public health research,[22] but from a proximate biological perspective, it is highly problematic and may obscure much of what it seeks to illuminate. The United States' history of slavery and segregation led to binary categorizations of race, including what became known as the "one-drop" or hypodescent rule, meaning that a single black ancestor defined a person as black. In contrast, many other countries have historically applied nuanced and graded racial categorizations.[23] Furthermore, genes for skin color are not linked to genes for health, intelligence, or other traits that have been culturally assumed to be related to race,[24] yet the US

National Institutes of Health requires researchers to use racial categories in most medical research, because epidemiologically race correlates with many health outcomes. But, as illustrated in Chapter 2, correlation does not equal causation. While underlying population variation in genetic predispositions correlates with health risks, most are not tightly linked to genes determining skin color. Rather, as Gravlee has elegantly shown, it is the experience of racism (which is stronger with darker skin) that becomes embodied that explains much of the correlation.[25] The biological effects of the social stress of racism are key to understanding the correlations. Thus, race is not a biologically useful construct, but rather a social construct with biological impact.[26]

The insights that come from studying calcium intakes and osteoporosis rates within and between populations may have important lessons for cross-cultural studies of aging and menopause. Relationships that seem to exist within certain populations are sometimes found to be more nuanced or even opposing when cross-cultural comparisons are conducted. The idea that calcium derived from dairy products is essential for preventing osteoporosis may hold true for US whites of European ancestry but may not hold true for people of different ancestry, or for those living in different environments. Cross-cultural perspectives may help us identify alternatives paths to similar outcomes. Could the example of osteoporosis and calcium hold any lessons for menopause and estrogen? Often, when we broaden our perspectives through cross-cultural comparisons, we discover that what we assumed to be "biological universals" are not so universal after all. Next, we turn to investigate the proximate biology of menopause, first examining the presumed textbook universal.

References

1. Räsänen J. 2021. Does the distinction between biological and chronological age support legal age change? *News*. Available at: https://bioethics.hms.harvard.edu/journal/legal-age-change#_edn1. Accessed 8 Jan, 2025.
2. Weon BM, Je JH. 2009. Theoretical estimation of maximum human lifespan. *Biogerontology*. 10(1):65–71. https://doi.org/10.1007/s10522-008-9156-4
3. Saunders T. 2024. The oldest person in the world in 2025. *The Human Body*. Available at: https://www.sciencefocus.com/the-human-body/oldest-person-in-the-world. Accessed 8 Jan, 2025.
4. He H, Xie B, Xie L. 2018. Male pattern baldness and incidence of prostate cancer: A systematic review and meta-analysis. *Medicine (Baltimore)*. 97(28): e11379. https://doi.org/10.1097/md.0000000000011379
5. Lotufo PA, Chae CU et al. 2000. Male pattern baldness and coronary heart disease: The Physicians' Health Study. *Arch Intern Med*. 160(2):165–171. https://doi.org/10.1001/archinte.160.2.165
6. Lu J, Shin Y et al. 2016. Peak bone mass and patterns of change in total bone mineral density and bone mineral contents from childhood into young adulthood. *J Clin Densitom*. 19(2):180–191. https://doi.org/10.1016/j.jocd.2014.08.001

7. Chevalley T, Rizzoli R. 2022. Acquisition of peak bone mass. *Best Pract Res Clin Endocrinol Metab*. 36(2):101616. https://doi.org/10.1016/j.beem.2022.101616
8. AFAR. *Theories of Aging*. 2016. Available at: https://www.afar.org/imported/AFAR_INFOAGING_GUIDE_THEORIES_OF_AGING_2016.pdf
9. Passarino G, De Rango F, Montesanto A. 2016. Human longevity: Genetics or lifestyle? It takes two to tango. *Immun Ageing*. 13(1):12. https://doi.org/10.1186/s12979-016-0066-z
10. Wang K, Liu H et al. 2022. Epigenetic regulation of aging: Implications for interventions of aging and diseases. *Signal Transduct Target Ther*. 7(1):374. https://doi.org/10.1038/s41392-022-01211-8
11. Pal S, Tyler JK. 2016. Epigenetics and aging. *Sci Adv*. 2(7):e1600584. https://doi.org/10.1126/sciadv.1600584
12. Levine ME, Lu AT et al. 2018. An epigenetic biomarker of aging for lifespan and healthspan. *Aging (Albany NY)*. 10(4):573–591. https://doi.org/10.18632/aging.101414
13. O'Toole PW, Jeffery IB. 2015. Gut microbiota and aging. *Science*. 350(6265):1214–1215. https://doi.org/10.1126/science.aac8469
14. Parker A, Romano S et al. 2022. Fecal microbiota transfer between young and aged mice reverses hallmarks of the aging gut, eye, and brain. *Microbiome*. 10(1):68. https://doi.org/10.1186/s40168-022-01243-w
15. Zhang WB, Ye K et al. 2021. The antagonistic pleiotropy of insulin-like growth factor 1. *Aging Cell*. 20(9):e13443. https://doi.org/10.1111/acel.13443
16. Pifferi F, Aujard F. 2019. Caloric restriction, longevity and aging: Recent contributions from human and non-human primate studies. *Prog Neuro-Psychopharmacol Biol Psychiatry*. 95:109702. https://doi.org/10.1016/j.pnpbp.2019.109702
17. Whyte HE, Findlay S et al. 2004. Dieting in adolescence. *Paediatr Child Health*. 9(7):487–503. https://doi.org/10.1093/pch/9.7.487
18. Flanagan EW, Most J et al. 2020. Calorie restriction and aging in humans. *Annu Rev Nutr*. 40:105–133. https://doi.org/10.1146/annurev-nutr-122319-034601
19. Beetch M, Harandi-Zadeh S et al. 2020. Dietary antioxidants remodel DNA methylation patterns in chronic disease. *Br J Pharmacol*. 177(6):1382–1408. https://doi.org/10.1111/bph.14888
20. Wiley AS. 2016. *Re-Imagining Milk*. New York: Routledge.
21. Shlisky J, Mandlik R et al. 2022. Calcium deficiency worldwide: Prevalence of inadequate intakes and associated health outcomes. *Ann NY Acad Sci*. 1512(1):10–28. https://doi.org/10.1111/nyas.14758
22. Lewis C, Cohen PR j 2023. Race and ethnic categories: A brief review of global terms and nomenclature. *Cureus*. 15(7):e41253. https://doi.org/10.7759/cureus.41253
23. Davenport L. 2020. The fluidity of racial classifications. *Annu Rev Political Sci*. 23:221–240. https://doi.org/10.1146/annurev-polisci-060418-042801
24. Jablonski NG. 2021. Skin color and race. *Am J Phys Anthropol*. 175(2):437–447. https://doi.org/10.1002/ajpa.24200
25. Gravlee CC. 2009. How race becomes biology: Embodiment of social inequality. *Am J Phys Anthropol*. 139(1):47–57. https://doi.org/10.1002/ajpa.20983
26. Goodman AH, Moses YT, Jones JL. 2019. *Race: Are We So Different?* Hoboken, NJ: Wiley Blackwell Available at: https://understandingrace.org/

4

PROXIMATE BIOLOGY OF MENOPAUSE

Menopause provides a powerful biocultural window into aging and is a part of life history that remains poorly understood. As population aging worldwide becomes a public health concern, female reproductive aging and menopause have received increasing attention because many aspects of women's health change during menopause.

Menopause is a natural event that encompasses the end of a woman's ability to biologically produce children (fecundity) and which all people with ovaries experience if they live long enough, although the timing, duration, endocrinology, and symptomatology of the transition appear to be variable. The mean age at menopause (or last reported menses) ranges from 45 to 51 years in population reports,[1] with a median range of 43.0–51.4 reported in populations studied between 1934 and 1982.[2]

In this chapter, we start with symptoms and signs of menopause, because they are what often lead women to seek healthcare and treatments, and where medicine has focused on trying to understand impacts on quality of life and health. In terms of proximate explanations of menopause, much focus has been on hormone changes, particularly changes in estrogen.

To make sure we are all on the same page, before diving into a discussion of menopause, we will review what has been canonized as "normal" premenopausal reproductive functioning.

4.1 Biology of "Normal" (Premenopausal) Functioning

There is a classic figure (Figure 4.1) of the "normal" menstrual cycle from a medical textbook chapter entitled "Disorders of the ovary and female reproductive tract" that shows changes in hormones, ovarian histology,

DOI: 10.4324/9781003644521-5

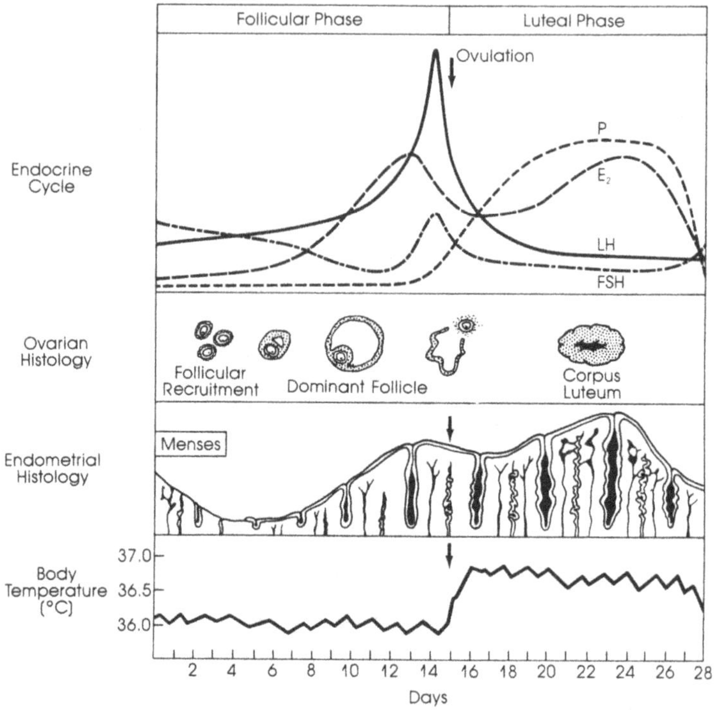

FIGURE 4.1 The "normal" monthly menstrual cycle.[3] Each panel describes changes throughout the cycle, which is divided by ovulation into two phases, follicular and luteal. From top to bottom panels, changes occur in: hormones including progesterone (P), estradiol (E2), luteinizing hormone (LH), and follicle-stimulating hormone (FSH); cellular level in ovarian follicles; lining of the uterus; and core body temperature. (reproduced with permission)

endometrial histology, and body temperature in a presumably average menstrual cycle.[3] This is considered the "textbook normal" 28-day menstrual cycle, which is divided into two phases: the follicular phase which leads up to ovulation (release of an egg), followed by the luteal phase. The latter phase differs if fertilization occurs, and thus this figure (and its title) seems to imply that fertilization and pregnancy may not be considered "normal."

There is a great deal of information in Figure 4.1. In the top panel (endocrine cycle), we see patterns of various hormones, without any units on the axis, which is interesting given the very different concentrations of these hormones. Also, follicle-stimulating hormone (FSH) and luteinizing hormone (LH) are released in a pulsatile fashion, which is not captured at this time scale. FSH stimulates egg-containing vesicles (follicles) to grow and to produce the hormones estradiol (E2) and progesterone (P). LH triggers an egg to be released (ovulation). After ovulation, the ovarian cells that form a temporary endocrine organ after ovulation (corpus luteum) produce progesterone that builds up the uterine lining to support potential pregnancy. If implantation does not occur (as in this figure), then progesterone decreases, and the uterine lining is shed in the menstrual period.

In the second panel, we see how the ovary changes on a tissue (histology) level. The third panel describes what happens with the uterine lining (endometrium), which when shed leads to menstruation. You can see the buildup of the lining in preparing for implantation of a fertilized egg. But if that doesn't happen, the uterine lining is shed as menses in a "normal" cycle. In the lowest panel of Figure 4.1, you can see how body temperature shifts over the cycle with an increase of about 1°C at ovulation, often used by women to assess risk or opportunity for pregnancy (basal body temperature or rhythm method for "natural" family planning). Progesterone, which increases after ovulation, acts on the hypothalamus to set higher basal body temperatures, so the luteal phase is often associated with increased feelings of warmth and perspiration. It is interesting that estrogen has been the primary focus of studies and treatments of menopausal symptoms like hot flushes, given the stronger association between progesterone and temperature in the "normal" menstrual cycle.[1]

(Note: I use hot flush [a term commonly used in the United Kingdom] instead of hot flash, to distinguish hot flushes from the foreign loan word *hotto furasshu* [hot flash] in Japanese.)

So, how do we define "normal"? For whom, at what age, in what population, and under what dietary and physical exercise conditions? The textbook "normal" menstrual cycle does not specify population, time of life, or energetic demands. This is typical of medical textbooks, and anthropologists love to critique these things (for good reason).

In the figure, the word "normal" is not defined. It is often used as a modifier for menstrual cycles that reflect what we assume to represent statistically average (center of the normal curve in a population distribution) hormonal values. Hormones change over the menstrual cycle, and at any time point, there is likely to be a normal distribution of hormone levels in any population. However, many key hormones of the hypothalamic-pituitary-gonadal axis, including gonadotropin releasing hormone (GnRH), FSH, and LH are released in pulsatile fashion,[5] making any snapshot of hormone levels somewhat problematic for estimating "normal."

But the word "normal" when referring to menstrual cycles also has the connotation of *normative*, referring to what we think *should* be normal. And this often has a WEIRD (Western, Educated, Industrialized, Rich, Democratic) bias, due to the populations that are studied. Language can be a powerful influence on our understanding of biology, as Emily Martin described in "The Egg and the Sperm: How Science has constructed a romance based on stereotypical male and female roles."[6] But how can one assess whether an individual (or even population) average menstrual cycle be considered "normal" without life course context (e.g., puberty, postpartum, menopause)? We know that hormonal levels vary not only between individuals during their cycles and throughout their reproductive lives, but also between populations,[7] as explored in the field of reproductive ecology.

4.2 Biological Definitions of Menopause

Biological definitions and understanding of menopause derive from investigations focused on changes at three anatomical and physiological levels: menstrual pattern, cellular, and hormonal. While symptoms such as hot flushes are considered hallmark signs of menopause, biomedical diagnoses of menopause require more "objectively" measurable indicators. Menopause is widely believed to reflect a depletion of ovarian follicles, and is characterized by a decrease in estradiol and an increase in gonadotropin hormones, culminating in irregular menstrual cycles and cessation of menses, sometimes with symptoms.[8]

4.2.1 Menstrual Pattern Level

The classificatory schema of menopause used in many research studies, including my own, relies on self-reported menstrual patterns in the previous 12 months, labeling the premenopausal woman as one who has experienced no menstrual changes, the perimenopausal woman as one who has experienced some change in menstrual patterns, and the postmenopausal woman as one who has had no menstrual periods in the previous 12 months. This schema is limited, however, because menstrual patterns may provide misleading information about ovarian function. For example, menstrual patterns may be irregular or stop due to exercise, dietary stress, and pregnancy, and thus the assumption of regular menstrual cycles before menopause may be flawed. (See Part III.)

4.2.2 Cellular Level

The maximum number of ovarian follicles occurs at about five months in utero, with between ½ and 1 million follicles present at birth, declining until menopause, when follicle numbers drop below 1,000.[9] Most are lost by follicular atresia (process of degeneration and resorption before maturation and release of egg). Usually only a single follicle releases an oocyte (egg cell), after which the follicle becomes the corpus luteum. Age at menopause is thus influenced by initial stock of follicles and rates of follicular atresia over the life course, which are influenced by lifestyle and exposure to factors such as smoking that vary between individuals and populations.[10] Lifestyle factors that correlate with age at menopause may provide clues to potential mechanisms and interventions, affecting not only age, but also experience, of menopause.

4.2.3 Hormonal Level

Declines in the so-called "female hormone" estrogen have received most of the focus in Western biomedicine of menopause. There are several forms of estrogen, including: estradiol (or 17beta-estradiol), the strongest (in terms of binding affinity for estrogen receptors and functional activity) and the primary estrogen produced in the body before menopause; estriol, the primary estrogen during pregnancy; and estrone, the weakest endogenous estrogen, which is prevalent post-menopause.

All these estrogen molecules are capable of binding estrogen receptors (ERs), and leading directly or indirectly to DNA transcription and signaling events that have a cascade of effects on cells and tissues throughout the body, including development of primary and secondary sexual characteristics, and regulation of reproduction and the menstrual cycle in women, as well as bone density, brain function, cholesterol mobilization, and control of inflammation in both women and men.[11] While estradiol is often labeled a female hormone, it also plays important roles in males in modulating libido, erectile function, and spermatogenesis.[12]

Both estrogen and progesterone levels exhibit inverted U-shape curves over the reproductive life course, with post-menopause levels similar to puberty levels. So why is menopause characterized as having low estrogen that needs to be replaced to achieve health, when that level is "normal" for that reproductive stage? We don't talk about low levels before puberty! Physiological parallels exist between puberty and menopause, in terms of how long the process takes, and how an equilibrium is reached. They can also be thought of as mirror images with puberty involving the system coming online and menopause involving the system going offline. Thus, while puberty and menopause bookend the reproductive phase of the life course for women, the attitudes toward these transitions and treatments differ significantly.

Dr. Jerilynn Prior has challenged the idea of menopause as an estrogen deficiency disease saying,

> The big problem with estrogen deficiency as a symptom or diagnosis is that it is a catchall term that plays into a cultural notion that estrogen is what makes a woman a woman. I am now 77 years old and am 20 years into my own menopause. Yet, I robustly argue that I am not "estrogen deficient" Why? Because my low estrogen and progesterone levels are *normal* for my age. And menopause is a *normal* life phase.[13]

4.2.4 Mapping Levels of Menopausal Changes

The widely used biomedical criteria for diagnosing menopausal stages were articulated by the 2011 Stages of Reproductive Aging Workshop (STRAW)+10 workshop.[14] While menopause may commonly be diagnosed by physicians relying on hormone levels to confirm stage, most women will assess their menopause stage by menstrual cycle patterns and symptoms. Changes in menstrual cycles can be subtle and variable. And changes may not be noticed, particularly by women who never had "regular" periods, whether due to pregnancy, lactation, exercise and dietary stress, or physiological variability. Symptoms such as thermoregulatory dysregulation leading to hot flushes and night sweats (vasomotor symptoms), as well as other physical and psychological symptoms, may be noticed by some women and indicate menopausal stage changes to them. But vasomotor symptoms are considered primary for diagnosis and are most likely to occur in the late perimenopausal stage, and immediately following the final menstrual period.

Hormones may be considered more 'objective' than symptoms. However, despite years of characterizing menopause as an "estrogen deficiency disease," estradiol does not even appear in the STRAW+10 criteria![14] Why not? Perhaps because there is too much variation in estradiol in "snapshot" blood tests to be predictive, hormones are released in pulsatile fashion and fluctuate throughout the menstrual cycle, and are dependent on complex feedback loops?

During the perimenopause, as follicle number decreases, estradiol decreases to levels below that required to provide adequate feedback to "turn down" FSH. This results in increased FSH, which in turn can cause increased estradiol production, leading to highly variable patterns of estradiol and FSH in early perimenopause. Depending on the timing of measurement, someone in the menopausal transition may have low or high estradiol and high or low FSH. (See Part II where we juxtapose the WEIRD focus on decreasing estradiol and need for replacement to bring it back to "normal" with the Japanese focus on hormonal balance, where the understanding is that hormones stabilize at new levels through each significant life course transition or *kawarime*).

It is not uncommon for women to have menstrual cycles without ovulation or textbook "normal" hormones. But I observed the opposite pattern as well, with one woman having seemingly "normal" changes in reproductive hormones throughout some of her cycles despite no menstruation for many years.

Fieldnotes

One of my Japanese participants told me that her periods had stopped at the age of 37, following a car accident. She wondered if the cause was some internal anatomical impact or the psychological shock of the accident. Her periods never started again, and she was worried about osteoporosis, as she was thin and fine-boned to begin with.

While not a proponent of HRT for "normal menopause," I referred her to a gynecologist colleague of mine, and she did decide to go on HRT. (This raises potential ethical questions about when/whether anthropologists should intervene in their participants' lives. Our conversation and my referral happened at the end of the study, so it did not bias study results, but what if she had shared this with me earlier in the study?)

Interestingly, later when we began analyzing the collected weekly blood samples for hormones, we observed that she still had "normal" premenopausal hormonal cycling in several months of samples even though she wasn't menstruating. Her concern about osteoporosis derived from her assumption that her lack of periods meant lack of ovarian function. But her hormone data looked typical of a premenopausal cycle, albeit with no menstrual periods. Clearly, she was not a textbook case of hormones and menopause, but how many people are?

This fieldnote highlights the challenges of trying to understand what's going on with the body when we rely primarily on external symptoms, and the problems with "textbook normality." Also, levels of reproductive functioning (menstrual, cellular, and hormonal) do not always match. All people with ovaries will go through hormonal menopause if they live long enough, but not all will have symptoms (and certainly not the same symptoms). We tend to rely heavily on external signs and symptoms (such as menstruation) because that is what women experience. And while lack of periods often indicates lack of fertility, it is possible to have fertile hormonal cycles (for example after giving birth and breastfeeding) before menses resumes (since menses occurs at the end of a cycle where implantation of a fertilized egg has not occurred). And while lab tests can be run, they usually only identify when hormones are outside of "normal" ranges. Because the female reproductive system involves pulsatile release of hormones multiple times a day, and fluctuations throughout the menstrual cycle, it is very challenging to identify whether a woman's hormones might be contributing to symptoms if her hormones are in a broad "normal" range.

4.3 Signs and Symptoms of Menopause

4.3.1 Signs

Puberty and menopause are the bookends of female reproduction, characterized by changes in ovarian cells and functioning, hormones in the brain-ovarian system (i.e., hypothalamic-pituitary-gonadal axis), and external signs such as changes in menstruation. At puberty, the first menstrual period signifies the beginning of reproductive maturity and functioning (although the ability to carry a pregnancy to term may not happen for several years). At the opposite end of the reproductive life course, women cannot know that any particular menstrual period is their final one with certainty, until 12 months have passed, when they can be reasonably sure they will not have another one. Thus, people can only know they have experienced menopause in retrospect.

For most women, as they approach menopause, increasing irregularity in hormones is associated with irregular menstrual cycles, the most visible signs of changes near menopause. The hormonal changes are associated with symptoms such as vasomotor symptoms and vaginal dryness. Although all women will experience hormonal changes associated with menopause if they live long enough, not all women experience "hallmark" symptoms, suggesting that hormonal changes may be "necessary but not sufficient" or may be correlated with a yet unidentified causal factor.

Many people equate menopause with the cessation of menstruation. But many people may not realize that symptoms during the menopausal transition may also include extremely heavy periods, referred to as flooding.

Fieldnotes

One day, when I was working in the lab analyzing data on soy phytoestrogens from dried blood spots, the phone rang. When I answered the phone, the caller apologized for bothering me at work, but she sounded upset and concerned and said she wanted to ask me something. She was experiencing heavy menstrual bleeding, and wondered if that could be a symptom of menopause. I assured her that it could – and that it was called "flooding" in English. I didn't know the Japanese translation for that symptom, and fortunately she spoke some English. I hoped that just knowing that there was a word (whether in English or Japanese) for her symptoms might reassure her.

But she continued. She told me she was worried that she could be having a miscarriage – that the symptoms felt like that. I told her that heavy periods were certainly possible at menopause, but also told her that if it were possible that she might be having a miscarriage, she should see a doctor. This story highlights how potentially alarming changes in the perimenopause can be, especially if people are unprepared for them. It turned out that she went to the doctor and confirmed it was not a miscarriage, but rather "simply" perimenopausal flooding. Context is critical to how women respond and interpret such changes in their bodies, and information helps shape expectations about what is normal and what is a sign of something that needs medical attention.

4.3.2 Symptoms

In the West and Global North, the hallmark menopausal symptom discussed in the media and doctors' offices and the focus of many cartoons, is the hot flush. Hot flushes are classic vasomotor symptoms, so-called because they involve vasodilation (involving increased skin temperature and blood flow) to dissipate heat when women become overheated and begin to sweat and flush (usually from the head to the upper back-chest area).[15] The physiology is not well understood, but many systems seem to be involved in the complex thermoregulatory system.[16] Hot flushes are also not inevitable, nor ubiquitous, and maybe not even the most common symptom in all populations around the world.

Fieldnotes

Before beginning my research, I traveled to Tokyo in the summer to assess feasibility of my proposed study, and to identify potential field sites and collaborators. I usually took the subway, as that was the most efficient way of getting around the city. While the trains were air conditioned, walking through the labyrinth of tunnels connecting the various subway lines could often be quite hot and humid – partly due to the outside temperature, and partly due to the sheer numbers of people. I often found myself sweating – even when I was barely exerting myself.

But while sitting or standing on the trains I often noticed women who appeared to be wearing nylon stockings and sweaters, and they didn't seem to be sweating at all. I tried to dress up for these initial meetings but found that I was often drenched in sweat by the time I arrived. So, I took to wearing a t-shirt and light pants for traveling, and to packing nicer clothes in my backpack. I would arrive at my destination and bathe discreetly in a restroom – mopping the sweat from my face, neck, and torso and change into my more professional-looking clothes. I would change before each meeting, and then change back again to travel home. I discovered that all Japanese carry handkerchiefs with them – sometimes made of silk or cotton, but also of terrycloth. They use these to dry their hands after washing them. But also in the summer, they use them to dab at sweat on their faces. So, they did sweat – but compared to me, not often or profusely. This difference in apparent sweating response should have been my first clue to what I would find when I began my research – that Japanese women complain much less of hot flushes and sweats than of symptoms like chilliness, at the other end of the thermoregulatory continuum.

TABLE 4.1 Symptoms (not exhaustive) associated with menopause[17]

Symptom category	*Symptoms*
Vasomotor symptoms	Hot flushes; night sweats (hot flushes at night); temperature dysfunction, and constriction or dilation of blood vessels
Sexual symptoms	Vaginal dryness – often associated with painful intercourse; changes in libido; urological symptoms (e.g., incontinence, increased frequency of urination, urinary tract infections)
Psychological symptoms	Depression; anxiety; irritability; memory loss; difficulty concentrating; mood swings
Physical symptoms (non-vasomotor)	Headaches, which may start or end at menopause; aches and pains in various parts of the body; gastrointestinal difficulties; fatigue

There are many other symptoms associated with menopause and reproductive aging (see Table 4.1). Not all women have all of them; some women have none. Night sweats often cause sleeping problems, and that combined with loss of libido and vaginal dryness can impact sex life. Although some people joke that symptoms such as insomnia might not be such a problem if you could engage in entertaining activities like sex to pass the time.

Many ostensibly menopausal symptoms and hormonal changes also correlate with increasing age, posing a challenge to much research on aging and menopause.

Symptoms that are related to *changing* hormones of menopause would be expected to increase early in the menopausal transition and then decrease after menopause has occurred. For symptoms that increase at menopause but continue to increase after it is over, they may be more age-related, and possibly independent of, or not solely influenced by, menopause. One question is whether underlying hormonal changes of menopause accelerate age-related increases in symptoms.

It is often assumed that symptoms that occur around the age of 50 must be due to menopause, but because most studies lack adequate control populations (e.g., younger or older women, men), it is difficult to tease out symptoms that are due to underlying hormonal changes of menopause from those that might be related to aging in general. In Part II, I discuss a study that I conducted to compare midlife symptoms in women and in men.

4.3.3 Disease and Illness of Menopause

The symptoms of menopause are often viewed (in the West) as manifestations of an underlying "disease state" diagnosed by changes in menstrual patterns and hormones (e.g., decreasing estrogen, or increasing FSH). But there is no clear failproof biomarker that predicts which women will have which symptoms. Many women have concerns about increasing risks of osteoporosis and heart disease after menopause, but many other factors such as diet and physical activity influence those risks.

Should menopause be considered a disease or a natural part of life history, like puberty or pregnancy, which are frequently medicalized but not to the extent that menopause is in the West? While any "natural" life stage or transition can have challenges, it does not mean that all people experience them or have health risks. The illness experience of menopause is influenced by variation in cultural ideas and beliefs about women's roles (e.g., whether older women are revered or thought of as marginalized), the importance of reproduction, expectations about the body throughout the life course, and medicalization (discussed further in Part IV).

4.4 Mechanisms of Menopause

While many biocultural factors influence menopause, here we focus on proximate biological factors involved in reproductive aging.

4.4.1 Mechanisms of Reproductive Aging and Menopause

Genes may contribute to symptoms and experience of menopause, but data are limited. While several genes contribute to age at menopause, all together they explain very little (<3% in one large study) in observed variation,[18] suggesting that environmental factors such as smoking (or yet unknown genes) likely play the biggest role in influencing age at menopause.

In epidemiological studies with diverse US populations such as the Study of Women's Health Across the Nation (SWAN), potentially modifiable factors such as education, oral contraceptive use, weight, employment, smoking, alcohol, and physical activity have been observed to correlate with age of menopause, while race/ethnicity did not, after controlling for sociodemographic, lifestyle, and health factors.[19] A review of factors associated with menopausal symptoms in longitudinal population studies found ethnic, geographical, and individual factors affect symptom prevalence and severity, and in turn affect how symptoms are associated with other health-related conditions.[20]

All women, if they live long enough, will experience menopause-associated changes in hormones (decreased estradiol, increased FSH and LH), and stop menstruating. But not all women have symptoms like hot flushes, and for those who do, symptoms usually decline over time without any treatments. Epidemiological associations of hot flushes with lower estradiol suggest that hot flushes may be caused by lower estradiol. But decreased estradiol might be necessary but not sufficient for hot flushes, or correlated with some other causal factor (e.g., progesterone). The complex phenomenon of menopause provides ample food for thought in teasing apart correlation from causation (Figure 4.2).

4.4.2 Correlation vs. Causation in Menopause

Much of Western biomedicine has assumed that because hot flushes seem to occur during perimenopause when estradiol levels decrease, therefore decreases in estradiol must cause hot flushes. This has led to the idea of estrogen (or hormone) replacement therapy being the treatment of choice for menopausal hot flushes. But there are many other hormonal changes occurring during the menopausal transition, and it is possible that estradiol decrease is inversely correlated with hot flush occurrence but is not causal or that it is necessary but not sufficient to cause hot flushes. While it could be that decreased estradiol causes hot flushes, several other hypotheses deserve attention (Figure 4.2). Furthermore, dietary phytoestrogens have been hypothesized to act like estrogen, but there may be alternative explanations as well, related to factors correlated with soy intake.

While not shown in Figure 4.2, life course and developmental factors may play a role in the relationship between estradiol and hot flushes. Furthermore, premenopausal exposures (e.g., to oral contraceptives) or lack of exposures to pregnancy and lactation at evolutionarily expected times (see Part III) could predispose the system to HFs later in life. And to the extent that later life phytoestrogen intake correlates with earlier life phytoestrogen intake, it may actually be the earlier exposures that play a role in creating different thermoregulatory setpoints. Thus, earlier life course exposure could be the primary factor, or might be a necessary but not sufficient factor.

4.5 Treatment and Prevention of Menopausal Symptoms and Health Risks: Hormone Replacement Therapy (HRT)

People sometimes talk about menopause being like Puberty 2.0, but we treat them very differently. With puberty, we recognize and accept that hormonal changes and symptoms are "normal." So why not for menopause? Despite the reproductive system being a complex multi-hormonal system, biomedicine has focused attention and treatment almost exclusively on estrogen during the menopausal transition. If we think of menopause as an estrogen deficiency disease, we conclude that replacing it should alleviate symptoms and disease risks. But how much estrogen? To return to what hormonal phase of a long and changing reproductive life course? What exactly is the "normal" that is being sought? And other hormones (e.g., progesterone) also change at menopause, so why the focus on estrogen?

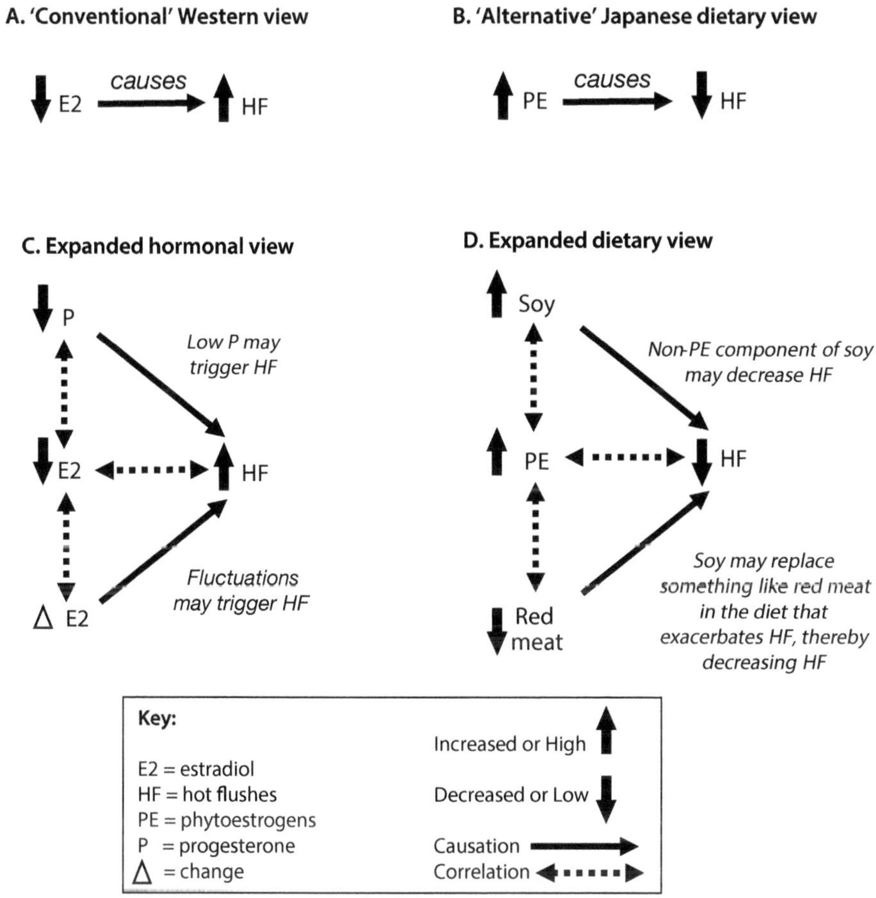

FIGURE 4.2 Possible causal vs. correlational relationships between menopausal hot flushes (HF) and endogenous hormones such as estradiol (E2) and progesterone (P), and exogenous hormones such as dietary soy phytoestrogens (PE). (a) "Conventional" Western biomedical view: decreased or low estradiol *causes* increased risk of HF. So, the prescription is to "replace" E2, causing HF to decrease. (b) "Alternative" Japanese dietary view: high PEs in the diet may cause decreased rates of HFs. (c) An expanded hormonal view considers possible correlations during the perimenopause: decreased or low E2 is correlated with decreased P, which could also trigger HFs, as could fluctuations and the rate of change in E2 and other hormones. (d) An expanded dietary view considers possible correlations with dietary factors: high soy PE intake is correlated with other soy components which may influence physiology of HFs; high soy intake is correlated with low intake of red meat (displaces it as a protein source) so it is possible that it is the absence of something in red meat that contributes to lower rates of HFs.

4.5.1 What Is HRT?

HRT is the use of extracted or synthesized hormones to replace endogenous hormones that have been lost through natural menopause or following removal of ovaries or with chemotherapy. HRT has a long[21] and controversial[22] history. The idea that menopausal women are lacking a crucial substance dates back at least to the 1800s when cow ovarian tissue was injected into women in Germany in order to treat sexual symptoms of menopause.[21] HRT became widespread once hormones could be administered orally and cheaply. Commercial preparations such as Premarin (estrogens from *preg*nant *mar*es' ur*ine*), accelerated interest in using hormones to treat menopausal symptoms.[23] Interestingly, it was later discovered that stallions have even higher concentrations in their urine. What does the fact that male horses have higher quantities of so-called "female" sex hormones mean? This highlights how culture and history influence how we view biological molecules, with implications for our ideas of biological sex and gender.

4.5.2 Cultural Context of HRT

Part 1 of this book is intended to cover proximate biology, while Part II examines proximate culture. Yet this division implies that what is presented in Part I is somehow not influenced by culture. The stereotype of biomedicine is that it is objective and universal. But we know that science is influenced by the beliefs and biases of scientists, which are in turn influenced by the culture and environment they live in. Thus, context and history are critical to understanding the assumptions and conclusions of science and medicine.

Foxcroft describes a European history, starting with the ancients, and focused on the United Kingdom and France, of the cultural context for attitudes toward menopause, and connections made between hormones and sex.[24] Lock wrote a historical analysis of attitudes toward aging and menopause in Japan and North America, elucidating where the hyper focus on estrogen came from.[25] The US history linking hormones with sex provides an important backdrop for understanding the history of HRT, the use of which increased in the 1960s, peaked in popularity in the 1990s, and then declined after results of clinical trials suggested it was not the panacea many people believed.[22]

In the 1960s, many cultural changes influenced interest in exogenous hormones. The birth control pill became widely available, and researchers Masters and Johnson published the best-selling *Human Sexual Response* about human physiology and anatomy. Perhaps one of the most influential events for HRT was gynecologist Dr. Wilson's publication in 1966, of the book, *Feminine Forever*. The cover reads: "In this first, fully documented book for women, a leading doctor in the field of menopause prevention explains why no woman – no matter what her age – need ever feel a day over forty."[26] This suggests that the physiology of a woman in her 30s was the goal of HRT.

A different cover for the same book states:

> The documented story of one of modern medicine's most revolutionary developments and breakthroughs – the realization that menopause is a hormone deficiency and totally preventable. Now, almost every woman, regardless of age, can safely live a full sex life for her entire life.

So, estrogen = female = sex life forever. This statement implies that "full" sex is not possible post-menopause. Do all women even want that? Some may, but certainly not all.

Just in case males are starting to feel left out, don't worry! Science and marketing have caught up and now offer testosterone replacement therapy for men and books to deal with andropause that promise a "happier, healthier and more sexually-satisfying life."[27] Does this sound similar to the rhetoric of *Feminine Forever*?

Most people think of testosterone as the "male hormone." One of the things that's fascinating about the way testosterone works in the body, and particularly in the brain, is that testosterone is often converted to estrogen locally by the enzyme aromatase.[28] Thus, the hormonal effects of the "male hormone" testosterone in the brain are often driven by what we call the "female hormone" estradiol. Perhaps our labels are problematic? If our ideas about sex hormones are highly oversimplified, could this interfere with our understanding, and bias toward oversimplified treatments and interventions for both women and men?

4.5.3 History of HRT and Chronic Disease

Early HRT primarily involved estrogen, approved by the FDA for treatment of hot flushes in 1942.[21] In the second half of the 20th century, observational studies suggested that HRT might prevent chronic diseases such as osteoporosis and cardiovascular disease, so the narrative shifted from "feminine forever" to "healthy forever."[22] Menopause does not cause heart disease directly, but changes associated with menopause such as decreased estrogen and increased cholesterol were hypothesized to play causal roles. With increasing postmenopausal life expectancy and aging populations, drugs treating chronic diseases with long prevention and treatment horizons have greater market potential than drugs treating hot flushes, which generally resolve after a few years.

Use of unopposed estrogen was associated with increased endometrial cancer risk, which was "solved" by decreasing estrogen and adding progesterone,[22] suggesting that the sole focus on estrogen was perhaps misguided and highlighting potential tradeoffs between cancer and other chronic diseases.

While HRT marketing shifted from a focus on hot flushes and sexual symptoms to prevention of chronic disease, and HRT prescriptions increased, clinical trials such as the Heart and Estrogen/Progestin Replacement Study (HERS) and Women's Health Initiative (WHI) began to examine the effects of HRT on chronic disease. Note that prescriptions increased before robust evidence of efficacy and safety were available! (Spoiler alert: at the time of writing, the FDA has still not approved any estrogen HRT for reduction of heart disease.)

4.5.4 WHI Study

Randomized placebo-controlled clinical trials are the gold standard method for assessing whether drugs really work the way we think they do. Participants are monitored closely for adverse events, and studies may be terminated early. In 1993, the HERS began recruiting and randomly assigning 2,763 postmenopausal women (aged 44–79 years, mean 66.7) with coronary heart disease to either an HRT or placebo group.[29] An increase in coronary events was found in year 1, but the study was not terminated early, and fewer coronary events were observed in years 4 and 5.[30]

The WHI study, initiated in 1992 with planned completion in 2007, was the largest longitudinal study designed to assess how HRT in menopausal women affected chronic disease risks.[31] There were four study arms: 1. Estrogen-plus-progestin (CEE/MPA), 2. Estrogen (CEE) only, 3. Diet (low fat/high fruits/vegetables/grains), and 4. Calcium and Vitamin D. Measured study outcomes were changes in risks (of intervention compared to placebo or control) of cardiovascular disease, stroke, osteoporosis fractures, colorectal cancer, and breast cancer.

The sample size, duration, and cost of the WHI study were unprecedented. But what was most significant about this study was that preliminary results indicating unexpected *increased* chronic disease risks associated with HRT led to terminating the WHI study early, a watershed moment in menopause research. Scientists (and funders) don't like to terminate studies early, and this only happens when study protocols pose significant risks to participants.

In 2002 (my first year of fieldwork), early WHI results showed that women who received estrogen-plus-progestin did not show the hypothesized decreased risks, but rather showed *increased* risks of coronary heart disease, stroke, and pulmonary embolism.[32] Also in 2002, after an average of 6.8 years of follow-up, the estrogen-only arm was also stopped because early results showed *increased* risk of stroke and *no protection* against heart disease.

A few years later in 2005, the dietary modification trial showed *no decreased* disease risk, counter to the hypotheses that a low-fat diet would reduce risk of breast and colorectal cancers, heart disease, and stroke in postmenopausal women. The Calcium-Vitamin D (CaD) Trial results showed only a modest increase in hip bone density, but *no risk reduction* for breast and colorectal cancers. Shockingly to most, study results did not simply fail to support hypotheses that HRT prevented disease, but actually suggested that HRT *increased* the risks of diseases it was currently being prescribed to prevent.

One might think that millions of taxpayer dollars were wasted – since the treatments did not have the positive results people predicted. Or we could consider this a great use of money, because it highlighted how treatments marketed to women on the assumption that they were beneficial were not only "not effective" but also "potentially harmful." Upon hearing the news, many women discontinued taking hormones immediately and researchers saw a dramatic increase in hot flushes because abruptly stopping hormones can trigger hot flushes. Until these results were published, many people believed women could take HRT safely until they died, not realizing that HRT might accelerate when that happened. Was it worth it for the youthful appearance and "sex life forever," even if the length of "forever" might be cut short by increased rates of cancer and heart disease?

Significant limitations of the WHI affected interpretation of results and implications for clinical practice, and these have been debated and subjected to re-analysis in subsequent years. The primary critique was that the average patient age on study entry was 63 years, meaning that the studies were stopped when two-thirds or more of these patients were 68 years or older. Thus, the average patient in their 40s and 50s using hormone therapy to alleviate menopausal symptoms is appreciably younger than the patients in the WHI study. Additionally, women with severe menopausal symptoms were excluded from WHI participation because possibly assigning women with severe symptoms to placebo seemed inappropriate. This means that study results are not relevant to women with severe symptoms, who constitute people most likely to consider HRT.

The WHI study was designed to test the hypothesis that long-term postmenopausal use of HRT would be protective against coronary heart disease and other chronic diseases, because that is what women were being told/sold. When data suggested long-term HRT might not actually be safe, prescriptions and marketing returned to the idea of HRT for treating menopausal symptoms, and there have been numerous re-analyses of the data to justify this use, since a reduced market for HRT is preferable to none. And of course, some women do have symptoms that interfere with quality of life, especially in a culture that expects women to function as they did in their 30s, and healthcare providers want to be able to offer women solutions.

In the 2020s, the North American Menopause Society position on hormone therapy is:

> For women aged younger than 60 years or who are within 10 years of menopause onset and have no contraindications, the benefit-risk ratio is favorable for treatment of bothersome vasomotor symptoms and prevention of bone loss. For women who initiate hormone therapy more than 10 years from menopause onset or who are aged older than 60 years, the benefit-risk ratio appears less favorable because of the greater absolute risks of coronary heart disease, stroke, venous thromboembolism, and dementia.[33]

How does one weigh the tradeoffs of treating bothersome symptoms and preventing bone loss with increasing risk of chronic disease? It seems like comparing apples and oranges, but with more important repercussions. Once a woman starts on HRT, how does she know when it is okay to stop, without triggering symptoms that led her to start it in the first place?

How was it that hormones were able to be prescribed and marketed for these conditions and symptoms without another large scale WHI-like study? What does it say about assumptions we make about the safety of "replacing" hormones during a period of life when they are "normally" at low levels?

Fieldnotes

In the 1990s I attended a conference presentation where the keynote speaker argued that because postmenopausal women experienced estrogen deficiency, every woman should be on HRT. Following his talk, a biological anthropologist posed a question. She said that she recognized that estrogen levels were low post-menopause compared to the reproductive years, but they also seemed low in early childhood and puberty. So, she asked if the speaker thought we should be giving HRT to adolescents and to children to make them "normal"?

The speaker was completely flabbergasted, and seemed to be trying to assess whether she was joking. She was trying to highlight the problematic logic of "replacing" something that wasn't expected. She was taking a broad life course perspective, and simply applying the logic to the other end of the reproductive lifespan. But in doing so, she caused the audience, and maybe even the speaker, to question a dearly held assumption about estrogen deficiency and HRT.

Few medical researchers and practitioners take a broad life course perspective when treating menopausal women. If postmenopausal women are considered estrogen deficient (arguably they are also progesterone deficient, and

TABLE 4.2 Contrasting views of menopause and HRT

Estrogen (E) deficiency disease – pro HRT	*Normal life course – no need for HRT*
• Disease risk increases after menopause, when E decreases (correlation) • Why not add E back into women's bodies, alleviating menopausal symptoms and decreasing risk of later life disease? • This could lead to healthier aging (for individuals and populations) and lower costs associated with osteoporosis, etc.	• Could increasing disease risk after menopause be influenced by lifestyle factors, that we should address first? • Why add E when it isn't expected from a life course perspective (e.g., post-menopause, or pre-puberty)? • Why not give E to males, who could be considered estrogen deficient, and also die at higher rates throughout the lifetime and have lower life expectancy?

have excess FSH), should we not consider preadolescent girls to be estrogen deficient, and thus treat both? Why does one proposition sound "normal" and one proposition sound ridiculous?

These contrasting views of menopause and HRT are shown in Table 4.2.

It is easy to say something can be used safely for the rest of your life when you haven't done any studies to assess whether it really is safe. This is what happened in the 1990s before the WHI. And it may be happening again in the 2020s, with advice that says it is safe to take HRT earlier and for shorter periods of time, even though no comparable large-scale WHI-like studies have been done to examine long-term effects of women initiating HRT around the age of menopause and using it for only a few years, or possibly many more. Another WHI-like study seems needed to ensure that history does not repeat itself with unintended consequences for millions of women.

Further complicating studies on whether, and to what extent, treatments for managing menopausal symptoms are effective is the placebo effect. To assess whether an intervention significantly alleviates symptoms, the scientific gold standard is placebo-controlled trials. In such studies, participants are randomly assigned to receive the drug or active chemical (e.g., HRT or specific components of soy) or a placebo (often called a sugar pill – where there is ostensibly no biologically active ingredient), and after a certain period of exposure, they are examined for differences in the outcome of interest. Drug companies hope to minimize the placebo response of their drugs, so that their drugs appear much more effective (when compared to placebo).

But I think the placebo effect should be harnessed, not minimized. Moerman has proposed that instead of the placebo response, we think of the "meaning response" of all health encounters (whether ingesting a pill, having surgery, or talking with a healthcare provider).[34] Often, when a person takes a placebo, if they think something's going to happen, something often does, showing the power of the mind and the meaning attributed to objects, behaviors, or care.

Placebo responses are often considerable in studies of menopausal hot flushes.[35] This might tempt some people to suggest that this means they are all in a woman's mind, which harkens back to sexist diagnoses of hysteria. However, as we increasingly understand the biocultural interplay between feelings and symptoms and recognize that many menopausal symptoms are heavily influenced by several brain regions, it makes sense to consider all interconnected aspects of this complex experience. Thus, women taking HRT or soy for relief of menopausal symptoms might receive not just benefits of bioactive chemicals but also benefits due to the "meaning response." The strength of the meaning response may tell us important things about the role of mind-body interconnections, particularly during a complex life course transition like menopause, and highlights the need for integration of biocultural perspectives for effective treatment.

4.6 Alternatives to HRT

Following the termination of the WHI study, many women as well as healthcare providers, decided hormones might not be worth the risk. Interest in complementary and alternative medicine (CAM) approaches to menopause increased, including phytoestrogens (plant estrogens) like those found in soy. CAM is commonly divided into two categories: mind-body practices and natural products (interesting since many pharmaceuticals derive from natural products). Mind-body practices, including cognitive behavioral therapy, hypnosis, biofeedback, mindfulness, and relaxation techniques may be effective in treating menopausal symptoms such as vasomotor, sexual, and sleep symptoms, and thus may improve quality of life.[36] By considering the role of the mind, we open up space for including proximate and ultimate cultural influences (see Parts II and IV) contributing to the local biology of menopause. Below, I focus primarily on dietary soy phytoestrogens because those may play an important proximate biology role in Japan.

4.6.1 Soy Phytoestrogens

> *Let food be thy medicine and let thy medicine be food.*
>
> —*Hippocrates*

This quote captures an idea still practiced in many Asian cultures. Soy foods have a long history in Japan, although considerable regional variation exists in the forms commonly eaten. In the United States, most people are familiar with tofu, and possibly soy protein supplements. But there is an incredible diversity of soy foods commonly consumed in Japan.

Fieldnotes

On my first trip to Japan, I was fortunate to find the Mitaka International Society for HOsPitality (MISHOP), a group of mostly middle-aged women, who met twice a week to study English with foreigners and teach them Japanese. I started attending their meetings and became friends with several of them. When I told them of my interest in soy, they threw a "daizu no pātī" (soy party) for me. Every person made a different soy dish – and there must have been about 20 people and delicious dishes including: tofu, yuba, miso *soup,* tempura nattō *(*nattō *wrapped in nori, battered, and fried),* edamame, age dashi dofu, okara *(soy pulp or tofu dregs that includes insoluble soybean parts remaining after pureeing soybeans in tofu making) in* unohana *(with carrots, burdock root, and shiitake mushrooms). This party was quite an event, showcasing the incredible range of forms in which soy is consumed in Japan. Many years later, whenever I would visit MISHOP, someone would inevitably mention the daizu party which assumed almost mythic proportions as the years passed!*

4.6.2 Soy Phytoestrogens: Structure and Function

While plants have been used as medicines for thousands of years, it wasn't until the 1990s when biological anthropologists began to study the significance of dietary sources of biologically active chemicals.[37–39] In contrast to xenoestrogens like BPA, DDT, and dioxin (which negatively impact environmental and human health), phytoestrogens appear to exert protective benefits on the reproductive system of both females and males. Phytoestrogens are naturally occurring plant chemicals with weak estrogenic properties due to structural similarities to estrogen[40] (Figure 4.3). The most common soy phytoestrogens are genistein and daidzein.

Even if you're not a chemist, hopefully you can see the structural similarities of these molecules. While estradiol has four rings, the soy phytoestrogens and metabolites have only three rings, but they are arranged in a similar 3D conformation to estradiol, so they are capable of binding estrogen receptors.

Fieldnotes

Soy has received much controversial attention. At a nutritional anthropology conference, I once received a question about whether soy could be exerting feminizing effects on men, and I jokingly responded, "Perhaps that is why the birth rate is so low in Japan?" The questioner initially seemed to think I was serious. I told him that I was joking, although there is evidence that other animals' reproductive cycles have been disrupted by phytoestrogens – for example, in animals like sheep that may eat diets containing high quantities of clover, which has high quantities of phyotestrogens.

I think it unlikely that Japanese men are consuming high enough quantities of soy that they are interfering with sperm production. Frequent alcohol consumption might be a more likely culprit. Also, there are many cultural factors such as later age at marriage, costs of raising children, women's reluctance to have children, and shortage of daycare facilities that may contribute to declining birth rates. And of course, sperm counts have been declining worldwide including in populations that do not eat soy, suggesting other environmental factors are at play. This story highlights the importance of considering alternative hypotheses for phenomena, and not seeing everything through the lens of a single factor like soy.

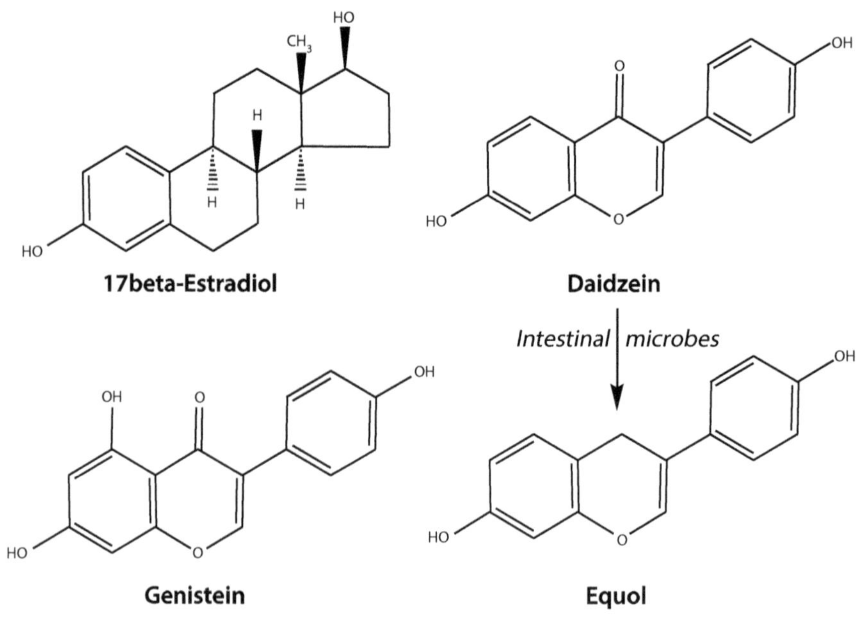

FIGURE 4.3 Similar chemical structures of primary endogenous estrogen (17beta-Estradiol), two soy phytoestrogens (Genistein and Daidzein), and one microbial metabolite of daidzein (Equol).

Most phytoestrogens in soy exist bound to other molecules. To be absorbed in the gastrointestinal tract, they must be broken down by enzymes, many of which are produced by gut microbes. Gut microbes not only break bonds between molecules and phytoestrogens, allowing them to be absorbed into the blood, but also convert phytoestrogens into important metabolites. These phytoestrogens and metabolites can influence sex hormone production, metabolism and biological activity, intracellular enzymes, protein synthesis, growth factor action, malignant cell proliferation, differentiation, cell adhesion, and angiogenesis.[41] Studies suggest that equol, a daidzein metabolite that is particularly bioactive (due to stronger binding affinity for estrogen receptors), may account for some of the individual and population differences in observed effects of soy phytoestrogens.[42] For example, twice as many Asians as Americans seem to harbor the microbes that can transform daidzein into equol,[43] possibly explaining why Japanese seem to gain more benefit from consuming soy.

4.6.3 Epidemiology of Soy

Epidemiological studies examining correlations between soy intake and many health outcomes[44] support the hypothesis that soy phytoestrogens might contribute to lower rates of vasomotor symptoms and chronic disease in Japanese women. Inverse correlations between soy intake and menopausal symptoms as well as chronic disease have been observed in prospective cohort studies.[45] While these results suggest phytoestrogens may act like estrogen, unlike with HRT, soy intake is correlated with lower risk of breast cancer in a dose-response fashion (i.e., more soy intake, lower risk of breast cancer).[46] Timing may be critical as breast tissue is most sensitive to dietary influences in childhood and adolescence, and thus early life soy intake might be most protective against breast cancer.[47] And for those who think soy may only impact females, soy intake is also associated with decreased prostate cancer risk in males.[48]

While inverse correlations between soy intake and health risks are intriguing, correlation does not prove causation. Many possible confounders exist. For example, individuals eating more soy may have fewer menopausal symptoms because of phytoestrogens in soy, something else in soy, other dietary components that are correlated with soy (like fish), or because of the absence of something inversely correlated with soy (like red meat) (Figure 4.2).

4.6.4 Methods

Various methods are employed to tease apart these different hypotheses. Food frequency questionnaires provide information on average diet, capturing seasonal food intake, and minimizing daily variation that may be due to work vs. weekend schedules. Dietary records capture more detailed dietary information – usually over a 24-hour period or several days. Both methods measure soy intake, but because of variation in bacterial and endogenous enzymes in the gut, the amount of phytoestrogens that reach the blood may vary between individuals, even when the same amount and form of soy is eaten. Thus, blood samples are needed to measure the phytoestrogen concentrations that the body actually "sees."[49] These three methods can allow triangulation to get a better sense of dietary influences, as each method has strengths and limitations.

However, blood samples only provide "snapshots" of chemicals in the blood, whether that be estrogen or phytoestrogens. In the case of estrogen, pulsatile release patterns and differences in concentrations over the menstrual cycle lead to a range of "normal" concentrations. In the case of phytoestrogens, concentrations vary depending on time since ingestion, the food matrix (what they were consumed with), and the presence of enzymes (from the host or gut microbes) that influence whether phytoestrogens get absorbed and in what form.

Timing of sampling is important, depending on what is of interest. In many studies, blood samples are taken in the morning (to be consistent and avoid the influence of circadian rhythms). But, depending on when soy was last consumed, morning concentrations may appear lower for people who regularly consume soy for breakfast instead of dinner.

4.6.5 Mechanisms of Soy Phytoestrogens

Let's examine potential mechanisms by which phytoestrogens may act in the body. Phytoestrogens can bind to estrogen receptors, but can exhibit both estrogenic effects (mimicking estrogen) and anti-estrogenic effects (competing with estrogen). Interestingly, while phytoestrogens act like estrogen in decreasing menopausal symptoms, they seem to exert anti-estrogenic effects for breast cancer. In this sense, phytoestrogens are not simply plant estrogens, but rather selective estrogen receptor modulators (SERMs),[50] meaning they have more nuanced and complicated effects on estrogen receptors. How might this work?

Phytoestrogens are considered weak estrogens. Many phytoestrogens do not bind as strongly as estradiol to estrogen receptors. But even genistein, which has a binding affinity comparable to estradiol for receptors, induces only about half the activity of estradiol.[51] Yet, this does not provide the full story of what goes on inside cells.

On a cellular level, the overall effect of phytoestrogens will depend on what else is in the cell. During perimenopausal hormone spikes (when estradiol is high), phytoestrogens may compete for binding sites and displace some estradiol, thus having anti-estrogenic effects. When estradiol is low (e.g., post-menopause), phytoestrogens may "fill in" for estradiol, thereby acting estrogenically. Thus, it is possible that high circulating phytoestrogen

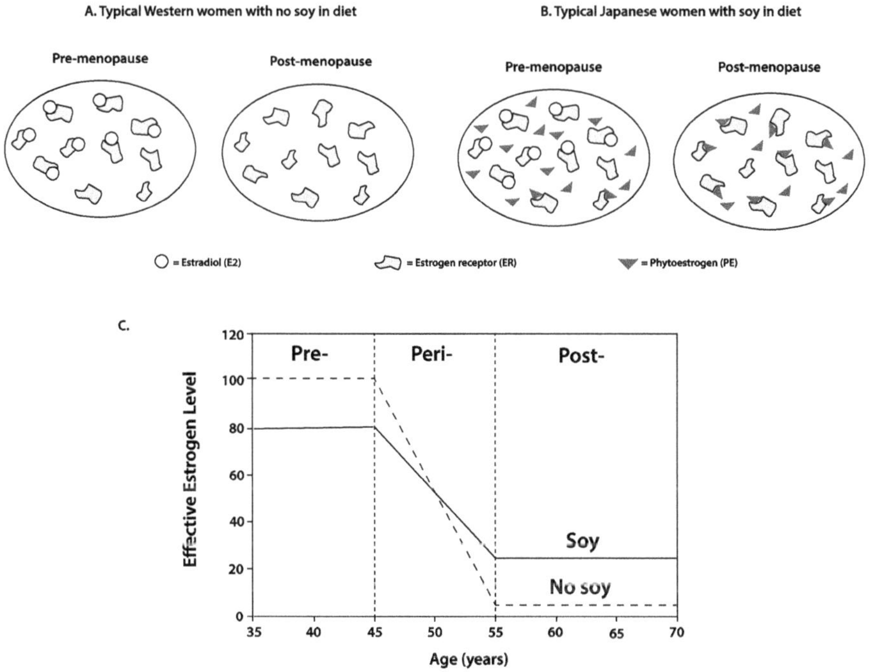

FIGURE 4.4 Schematic of hormonal environment of estrogen receptors pre- and post-menopause: (A) in typical Western women with no soy in diet and (B) in typical Japanese women with soy in diet. (C) Graphical representation comparing theoretically "effective estrogen" levels in women consuming soy and not consuming soy over the menopausal transition (note: the time scale of the x-axis is heuristic only; it could also be on the time scale of pulsatile endogenous hormone release), demonstrating the potential anti-estrogenic effects of soy phytoestrogens premenopause in the presence of estradiol, and estrogenic effects of soy phytoestrogens post-menopause in the absence of estradiol.

concentrations in the blood may buffer or balance out some of the fluctuations in estradiol on various time scales (Figure 4.4). And high concentrations obtained after ingestion of soy foods may compensate with quantity what they lack in quality (i.e., activity). Thus, in Japan where soy intake is high, significant effects of soy may be expected.

4.6.6 Food Processing by Microbes

Processing of soy by microbes, through both external fermentation and internal metabolism, may greatly impact not only the bioavailability and concentrations of soy compounds but also their forms. This demonstrates the importance of food processing as a cultural tool for modifying physiologically relevant aspects of the environment and creating local biologies.

The potentially significant physiological effect of this cultural practice in the Japanese context was reported from one study where intake of fermented soy products (*nattō* and *miso*), but not tofu, correlated with a decrease in frequency and intensity of hot flushes.[52]

Fieldnotes

Microbes play important roles in food processing inside and outside of the body. Many Japanese women make their own miso *(fermented soybean paste), which is used in soups and sauces for grilled vegetables, fish, or meat. Few people seem to make their own* nattō *– perhaps since it is harder to make. Or perhaps because* miso *is associated with the taste of* furosato *(one's hometown). I can appreciate the differences between red and white* miso*, and recognize that darker, saltier miso tends to predominate in the north and east of Japan (e.g., Fukushima), while the lighter white* miso *tends to predominate in the west (e.g., Kyoto). Regional differences in forms of soy consumed (e.g., fermented or not) may correlate with different exposures to microbes and their metabolites, and different health outcomes.*

The adage "you are what you eat" may require some modification, because much of what we eat may be food for our resident microbes, which then provide metabolites that are used to influence the working of host bodies. An expanded view of soy and food processing, both before (with fermentation) and after ingestion (with bacterial metabolism) suggests that soy and its constituents might more accurately be thought of as a substrate for microbes. The discovery that a microbial metabolite (equol) may explain some of the differential health effects of soy in some people[53] raises the possibility that microbes use soy substrates to produce the "real" active compounds. By eating soy, are humans "cultivating" certain bacteria, in a form of agriculture internal to the body? Where do those bacteria come from – especially those that may convert daidzein to equol? Studies have been done to see if eating soy could induce the ability to produce equol. Results have been mixed, and even non-equol producers seem to contain gut microbiota species that can convert daidzein into equol, although they do so at lower rates than in equol producers, for as yet unexplained reasons.[54] It has been suggested that key bacteria may reside in the intestines of fish, and thus the common practice in Japan of eating whole fish may lead to the relatively high proportions of equol producers in that population. This emphasizes the importance of expanding our view of relevant local biology and environment beyond just soy in the diet. These questions on microbial processing need to be investigated further, and I hope that readers of this book will be inspired to do so.

References

1. Melby MK, Lock M, Kaufert P. 2005. Culture and symptom reporting at menopause. *Hum Reprod Update.* 11(5):495–512. https://doi.org/10.1093/humupd/dmi018
2. Wood JW. 1994. Ch. 9 Menarche and menopause. In: *Dynamics of Human Reproduction: Biology, Biometry, Demography.* New York: Aldine De Gruyter; 401–439.
3. Carr B, Wilson J. 1987. Disorders of the ovary and female reproductive tract. In: Braunwald E, Isselbacher K, Petersdorf RG et al., eds. *Harrison's Principles of Internal Medicine.* 11th ed. New York: McGraw-Hill, 1987: 1818–1837.
4. Prior JC. 2011. Progesterone for symptomatic perimenopause treatment - Progesterone politics, physiology and potential for perimenopause. *Facts Views Vis Obgyn.* 3(2):109–120. Available at: https://pmc.ncbi.nlm.nih.gov/articles/PMC3987489/
5. Walker JJ, Terry JR et al. 2010. Encoding and decoding mechanisms of pulsatile hormone secretion. *J Neuroendocrinol.* 22(12):1226–1238. https://doi.org/10.1111/j.1365-2826.2010.02087.x
6. Martin E. 1991. The egg and the sperm: How science has constructed a romance based on stereotypical male-female roles. *Signs.* 16(3):485–501. Available at: https://www.jstor.org/stable/3174586
7. Vitzthum VJ. 2009. The ecology and evolutionary endocrinology of reproduction in the human female. *Am J Phys Anthropol.* 140(Suppl 49):95–136. https://doi.org/10.1002/ajpa.21195
8. Sievert LL. 2006. *Menopause: A Biocultural Perspective.* New Brunswick, NJ: Rutgers University Press.
9. Wallace WH, Kelsey TW. 2010. Human ovarian reserve from conception to the menopause. *PLoS One.* 5(1):e8772. https://doi.org/10.1371/journal.pone.0008772
10. Melby MK, Lampl M. 2011. Menopause, A biocultural perspective. *Ann Rev Anthropol.*40(1):53–70.https://doi.org/10.1146/annurev-anthro-081309-145641
11. Fuentes N, Silveyra P. 2019. Estrogen receptor signaling mechanisms. *Adv Protein Chem Struct Biol.* 116:135–170. https://doi.org/10.1016/bs.apcsb.2019.01.001
12. Schulster M, Bernie AM, Ramasamy R. 2016. The role of estradiol in male reproductive function. *Asian J Androl.* 18(3):435–440. https://doi.org/10.4103/1008-682x.173932
13. Prior J. 2021. Is estrogen deficiency really a thing? *Scientific American.* Available at: https://www.scientificamerican.com/article/is-estrogen-deficiency-really-a-thing/
14. Harlow SD, Gass M et al. 2012. Executive summary of the Stages of Reproductive Aging Workshop + 10: Addressing the unfinished agenda of staging reproductive aging. *Menopause.* 19(4):387–395. https://doi.org/10.1097/gme.0b013e31824d8f40
15. Deecher DC, Dorries K. 2007. Understanding the pathophysiology of vasomotor symptoms (hot flushes and night sweats) that occur in perimenopause, menopause, and postmenopause life stages. *Arch Womens Ment Health.* 10(6):247–257. https://doi.org/10.1007/s00737-007-0209-5

16. Freedman RR. 2014. Menopausal hot flashes: Mechanisms, endocrinology, treatment. *J Steroid Biochem Mol Biol.* 142:115–120. https://doi.org/10.1016/j.jsbmb.2013.08.010
17. Melby MK, Sievert LL et al. 2011. Overview of methods used in cross-cultural comparisons of menopausal symptoms and their determinants: Guidelines for Strengthening the Reporting of Menopause and Aging (STROMA) studies. *Maturitas.* 70(2):99–109. https://doi.org/10.1016/j.maturitas.2011.07.011
18. He C, Kraft P et al. 2009. Genome-wide association studies identify loci associated with age at menarche and age at natural menopause. *Nat Genet.* 41(6):724–728. https://doi.org/10.1038/ng.385
19. Gold EB, Crawford SL et al. 2013. Factors related to age at natural menopause: Longitudinal analyses from SWAN. *Am J Epidemiol.* 178(1):70–83. https://doi.org/10.1093/aje/kws421
20. Monteleone P, Mascagni G et al. 2018. Symptoms of menopause - Global prevalence, physiology and implications. *Nat Rev Endocrinol.* 14(4):199–215. https://doi.org/10.1038/nrendo.2017.180
21. Kohn GE, Rodriguez KM et al. 2019. The history of estrogen therapy. *Sex Med Rev.* 7(3):416–421. https://doi.org/10.1016/j.sxmr.2019.03.006
22. Cagnacci A, Venier M. 2019. The controversial history of hormone replacement therapy. *Medicina (Kaunas).* 55(9):602. https://doi.org/10.3390/medicina55090602
23. Vance DAD. 2007. Premarin: The intriguing history of a controverisal drug. *Int J Pharm Compd.* 11(4):282–286.
24. Foxcroft L. 2009. *Hot Flushes, Cold Science.* London: Granta Books.
25. Lock M. 1993. *Encounters with Aging: Mythologies of menopause in Japan and North America.* Berkeley, CA: University of California Press.
26. Wilson RA. 1966. *Feminine Forever.* Lippincott. New York, Pocket Books.
27. Howard B. 2014. *Andropause: The Complete Male Menopause Guide.* United States: Jett Publishing.
28. Gillies GE, McArthur S. 2010. Estrogen actions in the brain and the basis for differential action in men and women: A case for sex-specific medicines. *Pharmacol Rev.* 62(2):155–198. https://doi.org/10.1124/pr.109.002071
29. Grady D, Applegate W et al. 1998. Heart and Estrogen/progestin Replacement Study (HERS): Design, methods, and baseline characteristics. *Control Clin Trials.* 19(4):314–335. https://doi.org/10.1016/s0197-2456(98)00010-5
30. Hulley S, Grady D et al. 1998. Randomized trial of estrogen plus progestin for secondary prevention of coronary heart disease in postmenopausal women. Heart and Estrogen/progestin Replacement Study (HERS) Research Group. *JAMA.* 280(7):605–613. https://doi.org/10.1001/jama.280.7.605
31. The Women's Health Initiative Study Group. 1998. Design of the women's health initiative clinical trial and observational study. *Control Clin Trials.* 19(1):61–109. https://doi.org/10.1016/S0197-2456(97)00078-0
32. Rossouw JE, Anderson GL et al. 2002. Risks and benefits of estrogen plus progestin in healthy postmenopausal women: Principal results from the Women's Health Initiative randomized controlled trial. *JAMA.* 288(3):321–333. https://doi.org/10.1001/jama.288.3.321
33. NAMS Advisory Panel. 2022. The 2022 hormone therapy position statement of The North American Menopause Society. *Menopause.* 29(7):767–794. https://doi.org/10.1097/gme.0000000000002028

34. Moerman DE. 2002. *Meaning, Medicine and the 'Placebo Effect'*. Cambridge: Cambridge University Press.
35. Freeman EW, Ensrud KE et al. 2015. Placebo improvement in pharmacologic treatment of menopausal hot flashes: Time course, duration, and predictors. *Psychosom Med.* 77(2):167–175. https://doi.org/10.1097/psy.0000000000000143
36. Johnson A, Roberts L, Elkins G. 2019. Complementary and alternative medicine for menopause. *J Evid Based Integr Med.* 24:2515690x19829380. https://doi.org/10.1177/2515690x19829380
37. Etkin NL. 2006. *Edible Medicines: An Ethnopharmacology of Food*. Tucson, AZ: University of Arizona Press.
38. Jackson FLC. 1991. Secondary compounds in plants (allelochemicals) as promoters of human biological variability. *Ann Rev Anthropol.* 20:505–546.
39. Johns T. 1996. *The Origins of Human Diet and Medicine: Chemical Ecology*. Tucson, AZ: University of Arizona Press.
40. Gultekin E, Yildiz F. 2006. Introduction to phytoestrogens. In: Yildiz F, ed. *Phytoestrogens in Functional Foods*. Boca Raton, FL: CRC Press; 3–18.
41. Adlercreutz H, Mazur W. 1997. Phyto-oestrogens and western diseases. *Ann Med.* 29(2):95–120.
42. Mayo B, Vázquez L, Flórez AB. 2019. Equol: A bacterial metabolite from the daidzein isoflavone and its presumed beneficial health effects. *Nutrients.* 11(9):2231. https://doi.org/10.3390/nu11092231
43. Ortiz C, Manta B. 2024. Advances in equol production: Sustainable strategies for unlocking soy isoflavone benefits. *Results Chem.* 7:101288. https://doi.org/10.1016/j.rechem.2023.101288
44. Messina M, Duncan A et al. 2022. The health effects of soy: A reference guide for health professionals. *Front Nutr.* 9:97036.
45. Nagata C. 2021. Soy intake and chronic disease risk: Findings from prospective cohort studies in Japan. *Eur J Clin Nutr.* 75(6):890–901. https://doi.org/10.1038/s41430-020-00744-x
46. Wei Y, Lv J et al. 2020. Soy intake and breast cancer risk: A prospective study of 300,000 Chinese women and a dose-response meta-analysis. *Eur J Epidemiol.* 35(6):567–578. https://doi.org/10.1007/s10654-019-00585-4
47. Messina M, Hilakivi-Clarke L. 2009. Early intake appears to be the key to the proposed protective effects of soy intake against breast cancer. *Nutr Cancer.* 61(6):792–798. https://doi.org/10.1080/01635580903285015
48. Applegate CC, Rowles JL et al. 2018. Soy consumption and the risk of prostate cancer: An updated systematic review and meta-analysis. *Nutrients.* 10(1):40. https://doi.org/10.3390/nu10010040
49. Melby MK, Watanabe S et al. 2005. Sensitive high-performance liquid chromatographic method using coulometric electrode array detection for measurement of phytoestrogens in dried blood spots. *J Chromatogr B Analyt Technol Biomed Life Sci.* 826(1 2):81 90. https://doi.org/10.1016/j.jchromb.2005.08.009
50. Setchell KD. 2001. Soy isoflavones -- Benefits and risks from nature's selective estrogen receptor modulators (SERMs). *J Am Coll Nutr.* 20(5 Suppl):354S–362S; discussion 381S–383S. https://doi.org/10.1080/07315724.2001.10719168
51. Morito K, Hirose T et al. 2001. Interaction of phytoestrogens with estrogen receptors alpha and beta. *Biol Pharm Bull.* 24(4):351–356. https://doi.org/10.1248/bpb.24.351

52. Nagata C, Shimizu H et al. 1999. Hot flushes and other menopausal symptoms in relation to soy product intake in Japanese women. *Climacteric*. 2:6–12. https://doi.org/10.3109/13697139909025557
53. Lampe JW. 2009. Is equol the key to the efficacy of soy foods? *Am J Clin Nutr.* 89(5):1664s–1667s. https://doi.org/10.3945/ajcn.2009.26736T
54. Iino C, Shimoyama T et al. 2019. Daidzein intake is associated with equol producing status through an increase in the intestinal bacteria responsible for equol production. *Nutrients*. 11(2):433. https://doi.org/10.3390/nu11020433

PART II

Proximate Culture

5

PROXIMATE CULTURE

Introduction

In Part I, we examined physiological understandings of aging and menopause. Although these are often perceived as objective, Western biomedicine also derives from a particular historical and cultural context. This becomes clearer when contrasted with "alternative" conceptualizations of health, such as that in Japan. Because of the dominance of Western biomedicine and of English, alternative perspectives may be critical to gain a more holistic understanding of human experience.

What if non-Westerners (e.g., Japanese) had driven global medical science and Asian ideas were now globally dominant? Would Western perspectives on the body (e.g., body as machine and plumbing metaphors) be considered "alternative," while Asian perspectives on *ki* (energy) be considered standard and objective? How would we view aging differently in the West? Should any health/medical perspective ever be considered truly objective? To what extent are all health perspectives contextual or local?

Part II illustrates how difficult it is to place not only language, but lived experience, into simplistic categories humans create for themselves. While the concepts of aging and menopause appear more distinct in English and Western thought, they seem less so in Japanese language and thought. This is a great example of how culture influences what we often assume to be biological universals. And it is an organizational challenge for this book, which aims to integrate multiple perspectives and intersections, but is constrained by our linear structures.

Thus, in Part II, we examine how Japanese culture frames aging and menopause, by examining the concept and language of *kōnenki*, a word that is often translated as menopause but has a much broader meaning, perhaps more like the perimenopausal transition or climacteric. But *kōnenki* is not

DOI: 10.4324/9781003644521-7

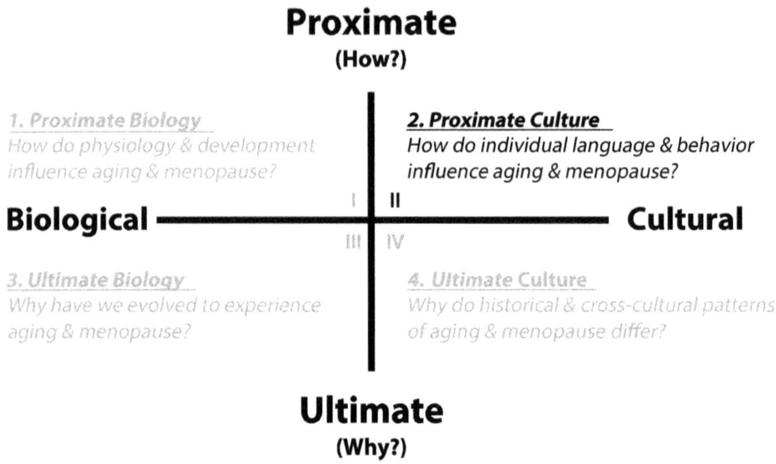

FIGURE 5.1 Proximate Culture Quadrant II of the proximate-ultimate biocultural framework introduced in Chapter 1.

restricted to menopausal changes per se or even women – as the word can also be used to refer to other transitions of the life course, and thus may be related to aging more generally.

The Chinese characters for *kōnenki* are: *kō* = renewal, regeneration, *nen* = years, and *ki* = season, period or stage. So *kōnenki* can be thought of as a period or stage of life of renewal or regeneration. And because Japanese is a symbolic language, when people see the word 更年期 (*kōnenki*), even if it has come to refer to menopause, it may also convey a sense of a season and time of change and renewal. Not simply a stopping of monthly cycles, as the English word menopause does.

Part II examines proximate culture (Quadrant II in Figure 5.1), and explores the question of *how* culture influences the experiences of aging and menopause, by examining language and behavior in Japan to better understand "insider" perspectives, implicitly comparing them with the "outsider" perspectives discussed in Part I.

5.1 Insider vs. Outsider Insights

We often think of biomedical researchers studying health in other cultures as having "objective" or "outsider" perspectives. The Western biomedical view of aging and menopause is described in Part I (Proximate Biology). But biomedicine is a product of particular historical and cultural contexts. When studying people from other cultures, anthropologists often try to "get into their heads" to ascertain the "subjective" or "insider" experiences. Then they try to convey this understanding to outside audiences using objective/outsider concepts and definitions. That is the focus of this part of the book.

Anthropologists use the terms *emic* and *etic*, derived from the linguistics terms phonemic and phonetic, to describe insider and outsider perspectives, respectively.[1] The insider/*emic* perspective focuses on the cultural distinctions that are meaningful to individuals of a given culture. Just as native speakers "know" if something sounds right in their language, native members of a culture are the best judges of the validity of an insider/*emic* description. The outsider/*etic* perspective relies upon extrinsic concepts and categories that have meaning for scientific observers and are believed to be culturally universal. For example, Western biomedical understandings of hot flushes as caused by decreased estrogen and as hallmark symptoms of menopause are considered objective and universal from the outsider/*etic* perspective valued by most Westerners. Scientists, and journal editors and reviewers, are the judges and gatekeepers of outsider/*etic* accounts of phenomena. As English has become the dominant language of science, most of these gatekeepers tend to have Western cultural biases.

But insider/*emic* and outsider/*etic* distinctions transcend simple language differences and extend to behavior. For example, in Japan, there is a concept of *amae*,[2] which describes the desire to depend on (or be indulged by) others, with the prototypical example being children's dependence on parents (verb form *amaeru*, meaning to depend on another's benevolence). Dependence, particularly in adulthood, is an insider/*emic* aspect of culture in Japan, especially when compared with a country like the United States where independence is highly valued. Examining insider/*emic* aspects of cultures expands our awareness of human variation and idiosyncrasies that exist between cultures.

Fieldnotes

*Since I am writing about Japanese language, even though I am not a linguist, I feel a need to try to demonstrate my "street cred" with the language. While my ability to read, and especially to hand write, Japanese characters (*kanji*), might call into question my authority (I have jokingly referred to myself as a functional illiterate in Japanese without a computer, at least in certain domains like politics, economics, and baseball), my ability to speak Japanese has probably made people think I am more fluent than I really am.*

*I was the only foreigner employed at the National Institute of Health and Nutrition at the time. When a new Director-General came in, replacing the one who had hired me (and who had been a long-term collaborator of mine and thus knew my strengths and weaknesses), he visited all the research labs and interviewed the directors and leaders of the various projects. I was the Food and Nutrition Education (*Shokuiku*) project leader at the time. One of the first questions he asked me was, "Melissa,* *how much Japanese can you read?" I was petrified he might give me some kind of test – such as ask me to read a newspaper that was lying on a nearby table. The adrenaline probably helped me come up with a clever, albeit panicked, response. I said,* "Sori-daijin to onaji gurai" *(about the same as the prime minister). Fortunately for me, the current prime minister, Taro Aso, had recently received a lot of press for misreading kanji when speaking publicly*[3] *– but few people questioned his intelligence or ability to do his job. Everyone around the table laughed at my very timely joke, and I passed the test, and got to keep my job as the only foreigner in the Institute. I considered sending a thank you note to the prime minister.*

Perhaps some languages are better at describing some parts of the world. Depending on context, certain parts of the social/physical environment may be more relevant and need more linguistic description (words for snow among Inuit being an oft-cited example). In Japan, words to convey social hierarchy or relationships with different levels of formality could be examples.

In science and health, many people trained in Western biomedical science believe that their worldviews about how things work are objective and universal. In the case of menopause, Western biomedicine has focused almost exclusively on estrogen as the cause of and treatment for menopause, despite other (even Western biomedical) voices arguing for an expanded view, including progesterone.[4] However, outside of the West, views may be even broader. For example, in Japanese explanations of menopause, the autonomic nervous system (ANS) plays a prominent role,[5] and a recent review lends support to this view.[6]

After reading Lock's book, *Encounters with Aging*, I was interested in how the Japanese insider/*emic* perspectives on *kōnenki* might map to Western biomedical outsider/*etic* understandings of what menopause and aging entail. Western biomedicine considers menopause a phenomenon caused primarily by decreased estrogen, with hallmark symptoms of hot flushes. Lock's book on menopause in Japan provided an insider/*emic* perspective that suggested significant gaps in insider-outsider perspectives about symptoms and experience of *kōnenki*-menopause. While menopause is often translated as *kōnenki* in Japanese, the terms and concepts differ.

To Westerners, other (non-Western) ways of knowing are generally considered non-objective, non-scientific, and thus *emic*. For example, the term "complementary and alternative medicine" (CAM) is often used without specifying the comparison – complementary and alternative to what? Usually to Western biomedicine. Thus, CAM often refers to Asian medicine in the Western biomedical literature. But for biomedically trained Japanese health practitioners who draw on a long tradition of Japanese *kampō* (herbal) medicine, which one should be labeled complementary or alternative? From a historical perspective, Western biomedicine is the more recent arrival to Japan, so perhaps biomedicine should be labeled alternative in Japan?

It is sometimes said that Western medicine treats the disease and Eastern medicine treats the whole person.[7] This makes randomized controlled trials of Eastern medicine challenging, since different people with the same disease may not be given the same therapies (Watanabe, pers. comm.) Eastern medicine uses many herbal treatments, which often contain bioactive compounds that may influence physiology. Western medicine also relies heavily on drugs derived from plants, even if most are extracts or synthesized compounds.

Cultural ideas on health and the body influence treatments. For example, Western perspectives and metaphors of the body as machine may influence how heart disease in the West is usually treated with surgery or "equipment" like stents to open or prevent blockages.[8] Such metaphors may contribute to overlooking psychological factors or lifestyle changes. Attitudes toward aging similarly use metaphors of body as machine, with solutions to replace damaged body parts (heart, knee, shoulder) or functions (ovarian hormones such as estrogen), just as we do with automobiles. How might different metaphors of the body influence the understanding and treatment of aging and menopause?

Fieldnotes

When I first met her, she was 75, the mother of one of my participants. She had bright eyes, set into a smiling face full of wrinkles, and a head covered with a scarf – all outward signs of her time spent in the sun, farming. When I commented that she looked so genki *(healthy), she replied (in a way typical of Japanese to downplay compliments) that she had bad knees, and the doctor was recommending knee replacements. But, as if for perspective, she added that none of her farming equipment ever lasted more than 25 years, so she thought she was doing great. This clearly shows some resistance to the Western biomedical views of aging and its solutions of replacing broken parts.*

5.2 Hypotheses for Japanese-North American Differences

What are the key elements of Japanese culture in the early 21st century that influence women's experience of aging and menopause? Lock's observations about midlife in Japan prompted a number of hypotheses regarding the mechanisms that might account for the differences between the experience of menopause in North America and in Japan including genetic, cultural, ecological, and physiological factors. Lock observed lower rates of hot flushes in Japan compared to North America, and suggested that could indicate "that women rarely have hot flushes, have them but do not notice them, or notice them but do not recognize them under the name given them by the researcher" (p. 498).[9]

So, how do we begin to tease these explanations apart? First, we must ensure that we inquire about symptoms using the correct linguistic term(s) to capture the experience. We might also try to measure hot flushes more objectively, by using a monitor to measure sternal skin conductance,[10] a physiological measure of hot flushes. Interestingly, there is not 100% concordance between objectively measured hot flushes and subjectively reported ones.[11] Sometimes people report hot flushes but the monitor does not record them perhaps because the threshold of the algorithm does not detect them, and sometimes the monitor detects/records them but people don't report them, suggesting a role for perception. The latter seems easily understood by people being distracted and not noticing or forgetting to report. But how do we explain the perception of a hot flush, without an objective measurement – could it be all in someone's head, or perhaps experienced somewhere else on the body besides the sternum?

Cultural values of individualism and collectivism are associated with functional MRI results indicating brain differences.[12] So, could culture also influence the experience and perception of hot flushes, and explain why there are differences in whether people subjectively notice "objectively" measured hot flushes?[10] For example, in one study, European-Americans were more likely than Japanese-Americans to report hot flushes that were not recorded objectively,[13] perhaps partly explaining observed population differences. Clearly, culture and biology are inextricably intertwined, despite my heuristic framework separating them in this book.

5.3 East Asian Medicine and Health Influences on Aging and Menopause

While Japan has highly developed biomedicine comparable to the best systems in the United States and Europe, it also has a highly developed system of traditional East Asian medicine.[14] Traditionally East Asian medicine views the body's ability to maintain stability or balance (homeostasis) holistically, resulting from dynamic interactions between *yin* (cold, heaven, night) and *yang* (warm, earth, day), incorporating physical, psychological, and social health, and thus inherently biocultural.

Utilizing the concept of *ki* (energy) to connect not only body parts, but also bodies with the broader environment, health and illness are seen as a continuum with processes that respond to environmental changes. While these traditional concepts still influence Japanese attitudes toward aging,

Western influence has played an increasingly important role. And interestingly, this connection between, or blurring of, the boundaries between the body and environment is increasingly being recognized in Western science with microbiome and epigenetics research, suggesting that the West might be a little late to this understanding.

Beginning in the late 1800s, Japanese medicine began to be influenced by German medical thinking, which introduced the concept of the interconnected autonomic nervous system, which easily aligned with the integrated Chinese-influenced medical system described above. More recently, Japanese biomedicine has been influenced by US biomedicine, particularly related to menopause and promotion of hormones. Despite high acceptance of biomedicine, the Japanese were initially reluctant to take birth control pills because they thought that hormone levels in US pills were too high for Japanese women's bodies. Reductions in estrogen amounts in pills over time from 150 to less than 50 micrograms suggests that initial levels were too high, not just for Japanese bodies but all bodies. Similar issues initially occurred with hormone replacement therapy (HRT), with many Japanese women expressing concern that HRT formulations prepared in the West would be too strong for Japanese women's bodies. Many participants told me that instead of "strong medicines" they preferred to try "gentler" traditional herbal medicine (*kampō*) and dietary changes first. And if possible, food and herbs are consumed to prevent illness in the first place.

The idea of food as medicine is often attributed to Hippocrates (460–377 BCE), but the idea of food therapy is found in earlier Chinese writings.[15] In East Asian medicine, tonic herbs, as well as acupuncture and other practices, are used to increase available energy to organs or parts of the body. These can be alternatives to biomedicine, or used complementarily, in conjunction with biomedical treatments. The existence of *kampō* clinics in hospitals such as Keio University Hospital demonstrates a greater degree of integration of biomedicine and "alternative" medicine than occurs in the United States.

East Asian ideas of health often integrate or blur the lines of nutrition and medicine, and many foods are believed to promote health. Most of my participants were very concerned with healthy eating and mentioned that they often ate more soy to make their blood be *sara-sara*, or flow easily (which they said they could observe when doing finger-prick blood spot sample collection for my research). They perceived that certain foods would make their blood thinner, and other foods would make their blood thicker or stickier. Several women also told me they felt constrained by having to cook food that their husbands and children would want to eat, and thus were not always able to eat as healthily as they would have liked.

In Japan, there is a strong emphasis on prevention, with high value placed on diet and eating seasonal foods to promote health: *sansai* (wild mountain vegetables) to promote health in spring, *unagi* (eel) to promote *genki* (vigor/health) in summer, *buri daikon* (yellowtail fish with Japanese radish) in winter, and fermented soy products like *nattō* and *miso* year-round. Blurring the lines between food and medicine (and traditional Eastern and more Western biomedical approaches), Japan actively promotes the use of food for health through the legal FOSHU (Food for specified health uses) designation,[16] including conditions relating to gastrointestinal health, cholesterol, blood sugar, and blood pressure. This reflects a more integrated medicine perspective, integrating Western and Asian medical approaches.

While dietary practices and exposures may be considered as proximate cultural factors with biological impacts, Japan has elevated its dietary culture to a societal and even global level after successfully petitioning for *washoku* (Japanese cuisine) to be designated as a UNESCO Intangible Cultural Heritage of Humanity, which appears linked to longevity in Japan.[17] This connects to ultimate culture described in Quadrant/Part IV.

The next two chapters examine aspects of culture and how they influence the local biology of aging and menopause in Japan. In Chapter 6, on aging, we explore the role of the autonomic nervous system, the concept of *gaman*, and environmental/lifestyle factors contributing to variation in the experiences and risks of aging. Then we will take a closer look at *kōnenki*, and how it may be similar to or different from aging and/or menopause, by examining *kōnenki* symptoms in control populations such as men, and the symptom of chilliness. We then turn to culture in the form of diet, exploring how soy phytoestrogens and their microbial metabolites might produce local biology in Japan, including a brief look at supplements.

In Chapter 7 on menopause, we ask whether *kōnenki* is equivalent to menopause, and explore how the language of hot flushes provides insights into underlying physiology and experience. We examine the potentially novel (to the West) vasomotor symptom of chilliness, which challenges many Western biomedical preconceptions about menopause. Finally, we examine links between vasomotor symptoms associated with menopause and soy intake and exposure.

The framework used in this book is heuristic and oversimplified. Part II, which explores *kōnenki*, is particularly problematic. The challenges I faced with trying to separate proximate cultural discussions of aging from menopause by examining *kōnenki* in Japan provide critical insights, just as the challenges of separating biology from culture led Lock to create the concept of local biologies (or situated biologies). Her work highlighted the need to think about the interactions between biology and culture. The framework and discussion that follow demonstrate the need to think about interactions between aging and menopause in a more nuanced way than is often thought of in the West.

References

1. Pike KL. 1954. *Language in Relation to a Unified Theory of the Structure of Human Behavior.* Dallas, TX: Summer Institute of Linguistics.
2. Doi T. 1973. *The Anatomy of Dependence.* Translated by John Bester. Tokyo: Kodansha International.
3. Yamaguchi M. 2009. Aso's kanji conundrums spur self-reflection, textbook sales. *The Japan Times*, 7 March.
4. Prior JC. 2011. Progesterone for symptomatic perimenopause treatment - Progesterone politics, physiology and potential for perimenopause. *Facts Views Vis Obgyn.* 3(2):109–120. Available at: https://pmc.ncbi.nlm.nih.gov/articles/PMC3987489/
5. Lock M. 1993. *Encounters with Aging: Mythologies of menopause in Japan and North America.* Berkeley: University of California Press.
6. Schwarz KG, Vicencio SC et al. 2024. Autonomic nervous system dysfunction throughout menopausal transition: A potential mechanism underpinning cardiovascular and cognitive alterations during female ageing. *J Physiol.* 602(2): 263–280. https://doi.org/10.1113/JP285126
7. Bhanote M. 2022. *Western vs. Eastern Medicine: What to Know.* Available at: https://www.medicalnewstoday.com/articles/western-vs-eastern-medicine. Accessed 23 Feb, 2022.
8. Payer L. 1988. *Medicine & Culture: Varieties of treatment in the United States, England, West Germany, and France.* New York: Henry Holt and Co.
9. Lock M, Kaufert P. 2001. Menopause, local biologies, and cultures of aging. *Am J Hum Biol.* 13(4):494–504. https://doi.org/10.1002/ajhb.1081
10. Freedman RR. 2001. Physiology of hot flashes. *Am J Hum Biol.* 13(4):453–464. https://doi.org/10.1002/ajhb.1077
11. Pachman DR, Loprinzi CL et al. 2013. Sternal skin conductance: A reasonable surrogate for hot flash measurement? *Menopause.* 20(11):1164–1168. https://doi.org/10.1097/GME.0b013e31828cec53
12. Chiao JY, Harada T et al. 2009. Neural basis of individualistic and collectivistic views of self. *Hum Brain Mapp.* 30(9):2813–2820. https://doi.org/10.1002/hbm.20707
13. Sievert LL, Reza A et al. 2010. Diurnal rhythm and concordance between objective and subjective hot flashes: The Hilo Women's Health study. *Menopause.* 17(3):471–479. https://doi.org/10.1097/gme.0b013e3181cbb3c6
14. Lock M. 1980. *East Asian Medicine in Urban Japan.* Berkeley: University of California Press.
15. Yao R, He C, Xiao P. 2023. 'Food and medicine continuum' in the east and west: Old tradition and current regulation. *Chin Herb Med.* 15(1):6–14. https://doi.org/10.1016/j.chmed.2022.12.002
16. Yamada K, Sato-Mito N et al. 2008. Health claim evidence requirements in Japan. *J Nutr.* 138(6):1192s–1198s. https://doi.org/10.1093/jn/138.6.1192S
17. Gabriel AS, Ninomiya K, Uneyama H. 2018. The role of the Japanese traditional diet in healthy and sustainable dietary patterns around the world. *Nutrients.* 10(2):173. https://doi.org/10.3390/nu10020173

6

PROXIMATE CULTURE OF AGING

Most people don't grow up. It's too damn difficult. What happens is most people get older. That's the truth of it. They honor their credit cards, they find parking spaces, they marry, they have the nerve to have children, but they don't grow up. Not really. They get older.[1]

—Maya Angelou

Angelou's quote highlights how cultural values and views related to aging may influence our perceptions and experiences of it. While contemporary Western countries like the United States seem to place high value on youth, other cultures like Japan have traditionally valued and respected elders for their experience and roles in society. Similarly, views on aging and life transitions differ between the West and Japan. Many of these differences are showcased in the nuance of the Japanese word *kōnenki*. A deep dive into the concept of *kōnenki* highlights the importance of culture in shaping the experience of aging.

In Japan, *kōnenki* is viewed as a natural life transition for women, and, depending on whom you consult, possibly even for men. While the English word "menopause" is often translated as *kōnenki*, the Japanese word *kōnenki* may be more accurately translated as **climacteric** (period of life when fertility declines, but also a critical period in the life course), a term that has fallen out of use, especially in North America. In Japan, there is an expectation that most women will experience *kōnenki* without medical intervention, like most other life transitions. Thus, while menopause is translated as *kōnenki* in Japanese, *kōnenki* may be more accurately thought of as a life stage or part of aging – but that is a claim we can try to assess empirically.

DOI: 10.4324/9781003644521-8

I wanted to know "What is *kōnenki*?" (insider/*emic* perspective) and "How does it differ from menopause?" (outsider/*etic* perspective).

6.1 East Asian Medicine Views on Aging

The concept of *kōnenki* reflects an East Asian understanding of health and the body. Lock described East Asian medicine with respect to healthcare for the elderly in Japan,[2] and documented the role of the autonomic nervous system (ANS), or *jiritsu shinkei*, in the understandings of the etiology of *kōnenki*.[3] The autonomic nervous system regulates involuntary physiological processes such as heart rate, blood pressure, respiration, digestion, and sexual arousal, and includes the sympathetic, parasympathetic, and enteric nervous systems. Links between the autonomic nervous system and endocrine system exist in the Japanese understanding of *kōnenki* – and may reflect a deeper, more holistic understanding of the physiological changes that occur at this time of life.

During my fieldwork, the term *jiritsu shinkei* seemed to have gone out of use, and women were more likely to give biomedical-sounding explanations for *kōnenki*, such as "hormone imbalance" (*horumon unbaransu*). Interestingly, they rarely spoke of "hormonal decrease" or "loss of estrogen" as is commonly said in North America, suggesting that the concept of imbalance deriving from East Asian medical understandings of the body and *jiritsu shinkei shitchōshō* (imbalance of the autonomic nervous system) have been blended with the North American biomedical focus on hormones. This idea of hormone imbalance during *kōnenki* implies an expectation that the body will return to balance again, even if it is a new balance or new "season" of life, characterized by different hormone levels. This contrasts with the idea of menopause as involving decreased estrogen, where the only solution is to add it back to the body, as detailed in Part I.

The concept of *gaman* (a Zen Buddhist term meaning to "bear the seemingly unbearable with dignity and patience") may shed light on the graceful aging of many Japanese women compared with their Western counterparts. *Gaman* also has implications for our ability to collect data on symptoms of aging and menopause, especially those that are bothersome and involve women's experience of pain or discomfort. In Japan, women's ability to endure hardships and continue taking care of their families is valued; therefore, women are likely to underreport severity of symptoms.[4] This Japanese concept of *gaman* resembles the British "stiff upper lip" stoicism, with an implicit assumption that if one has no choice but to persevere, why bother complaining about, or reporting, bothersome symptoms.

Many Japanese women reported they did not have the luxury of being able to indulge their symptoms of *kōnenki*, as they still had to take care of their husbands, children, in-laws, etc. Not until the caregiving phases of their lives were over, could they start paying attention to their symptoms, but at that point the practice of *gaman* might be so ingrained as to make it unlikely their behavior would change, or the recipients of their caregiving efforts might change from their children to their parents-in-law. Lock discussed the idea of being too *isogashī* (busy) to notice, or complain, about symptoms of *kōnenki*. If women are enculturated to *gaman*, will they complain or even accurately report bothersomeness or severity of symptoms – especially to physicians or researchers?

Fieldnotes

Participant observation often provides insights that interviews cannot. While I did not experience menopause myself when living in Japan, I did experience childbirth, twice, granting me insight into what my participants had shared about their own birthing experiences. Some women discussed how hospital stays following the birth of children could feel like vacations from all their duties at home (perhaps a well-deserved pampering after the gaman *required for childbirth).*

*Sometimes exceptions can prove the rule. When I gave birth in Japan and had severe chronic pain following the birth of my first child, Japanese doctors told me that foreigners were more sensitive and unable to endure (*gaman*) pain. This perception extended to giving birth without pain medication. Epidurals are hard to come by in Japan (only about 6% of women receive them) and many factors contribute to limited availability and use of analgesia during birth.*[5] *In contrast, epidural rates in the United States are more than 11 times higher, at greater than 70%.*[6] *Furthermore, many Western women who give birth in Tokyo seek out doctors that cater to expatriates and give epidurals.*

I hemorrhaged severely and was very anemic following the birth, and I was in chronic pain due to internal muscle tears that did not heal. Surgical repair and six months of pelvic floor therapy in the United States allowed me to function normally again.

When I returned to Japan and shared my story with Japanese women, several told me that they had experienced similar pain postpartum – that lasted for years or even decades. But they had never received help, and possibly not even mentioned it to their doctors, likely because of the cultural pressure to gaman. *I understood then that if women can live with such debilitating pain following childbirth, not complaining of menopausal hot flushes or other symptoms of aging was not surprising at all!*

Although many doctors told me at the beginning of my research that they thought it unlikely that women would ever share details about their symptoms with me, I found that women were eager to share details of their health and bodily functions. Perhaps this was because, as I was an outsider, they were not "complaining" to me but sharing details about their lives and experiences in response to my genuine interest. In addition, because I was 20–30 years younger than the people I was interviewing, there may have been some aspect of teaching me about what might be in my future.

In fact, in my experience, Japanese women seemed more open than most North Americans to reporting other bodily symptoms, perhaps a result of enduring influences of puritanism in US society. Many Japanese are not only concerned with what they put in their bodies, but also with what comes out (i.e., bowel movements as an indicator of health), reflecting East Asian medical attention to bodily signs and symptoms that many Americans would rather not discuss. Interestingly, in subsequent research on the microbiome in the United States, my colleagues and I found that many of our research subjects did look to gastrointestinal upset as signs of a disturbed microbiome that might need to be addressed, suggesting that things may be shifting in the United States.[7]

Cultural context is critical and the notion of *gaman* pervades Japanese culture, particularly when referring to women's bodies and experiences. Thus, the role of *gaman* must be considered in all studies involving symptom reporting. Even if Japanese women actually experience similar rates of symptoms compared to North American women, do they notice them, and do they report them?

Researchers such as Kleinman and Kirmayer have written extensively on somatization and cultural variation in somatic expressions of distress,[8,9] documenting that Asian cultures such as in China tend to have higher rates, relative to the United States, of somatization wherein psychological symptoms are expressed in terms of physical symptoms. In Japan, *gaman* influences not only how distress is expressed but whether it is conveyed at all. Thus, it is essential to examine inter- and intra-cultural variation in expressions of distress about somatic issues (ranging from exaggeration to *gaman* in the face of age-related symptoms).

6.2 *Kōnenki* as Climacteric of Aging?

Although often translated as menopause, the term *kōnenki* originally did not distinguish between men and women, and thus may be more accurately translated as "climacteric," which has a broader connotation of midlife transition.[10] Lock's book[3] documented how many Japanese people believe that women everywhere probably experience something like *kōnenki*, but also that *kōnenki* may be unique to Japan, in terms of its symptoms and the factors influencing its etiology. In lieu of a disease-bearing state, Lock's analysis brought the process of changing reproductive status into the realm of life history, more akin to the original meaning of *kōnenki* as a *kawarime* – a turning point in one's life – denoting a life cycle transition,[3] and thus perhaps more appropriately considered part of aging. Japanese also use the term *kōnenki shōgai*, with the latter word connoting disorders, or literally "obstacles" associated with *kōnenki*. These obstacles can be overcome as one transitions through the life stage, to find new balance and harmony on the other side. The word *kōnenki* illuminates how language can inform attitudes toward or acceptance of health experiences.

6.2.1 Kanazawa Study of Kōnenki: Men as "Aging Controls" for Menopause

How can we assess whether *kōnenki* is a natural part of aging or is equivalent to the Western biomedical phenomenon of menopause? One of the biggest challenges to understanding menopausal symptoms is that most studies of menopause and menopausal symptoms simply recruit women at midlife and ask them about certain symptoms. Then, they conclude that the symptoms that show high prevalence are menopausal symptoms, even though many of these symptoms may not be unique to, or even most prevalent in, menopausal women. Because studies of menopausal symptoms rarely include women outside typical perimenopausal ranges (ages 45–55 or 40–60), they do not collect data that would enable us to tease apart symptoms related to aging vs. menopause. One notable exception is a cross-cultural comparison of women aged 20–70 years old.[11] Interestingly, of the six symptom groups that were observed, only two were associated with hormonal changes of menopause: vasomotor symptoms (VMS) of hot flushes and night sweats; and poor memory, difficult sleeping, aches in head, neck and shoulders, vaginal dryness, and sexual arousal difficulties. Other symptom groups increased with age and body mass index, except for the group of psychological symptoms that peaked before age 40 and then declined.[11]

Symptoms that increase with age (and even those that decrease with age!) should be more accurately attributed to aging. Symptoms that peak during the menopausal transition, and then decline, may be more likely due to underlying (e.g., hormonal) changes of menopause. But if we only observe women during the menopausal transition, it is hard to differentiate among symptoms related to underlying causes of aging or menopause, or sociocultural factors associated with various life stages. We really need to compare younger and older women, and perhaps other "controls" such as men, who are aging but not undergoing such dramatic hormonal changes around the age of 50.

Interestingly, many of the health issues associated with menopause in the West, such as increased risk of heart disease and osteoporosis, are attributed to the loss of estrogen at menopause. Men have less estrogen to begin with, but also different heart disease and osteoporosis risks. In a culture such as Japan, where understandings of menopause are less about loss of estrogen, but rather about temporary imbalance, health risks may be viewed in a broader perspective and may yield insights beyond those studies focused on a single hormone like estrogen.

But first, which symptoms should be included on checklists distributed to study participants? The adage of "garbage in, garbage out" (GIGO), or flawed input produces flawed output, is relevant to study design and checklists. Inclusion, or omission, of a symptom in a checklist can compromise results. There is great variation in how many, and which, symptoms are included on commonly used menopausal symptom checklists. For example, the symptom of "formication" (feeling of ants crawling on one's skin), is found on the Blatt-Kuperman index, which was widely used in many studies of menopause in the latter half of the 20th century. It was created by a clinician in New York City in the 1950s, but few women would probably recognize the term today, and the evidence for it being a characteristic symptom of menopause is weak at best.[12] While inter-country comparisons are challenging due to language and cultural differences, checklists may vary considerably even within the same country, with varying implications and conclusions.[13]

To examine the extent to which common symptoms of *kōnenki* are unique to women, I conducted a study administering Lock's extensive two-week recall symptom checklist to 60 people, including women as well as men, of various ages in Kanazawa, Japan.[14] This was my first study on *kōnenki* in Japan, and I was excited to examine symptoms in both women and men, but unfortunately, I made a big error in my study design.

I thought I was being clever, administering a "menopausal symptom checklist" to both men and women with the goal of teasing out which symptoms were unique to women. However, in my excitement, and perhaps relative inexperience, I neglected to notice one important missing question on Lock's survey that I should have added for this study: gender (as proxy for biological sex)!

I sent questionnaires home with my fellow Japanese language students to be given to their homestay families, and I neglected to record the variable of gender – even as administering the questionnaire to both men and women was a primary study aim. Of course, Lock's survey did not have gender as a demographic question, because – like all other menopausal surveys – she was giving the questionnaire only to women! It is only when I started entering the data that I realized my mistake (much to my utter horror). I then had to send a second survey explaining my mistake and asking participants for their gender, as well as a few other identifying characteristics (e.g., age) so that I could match up the survey responses. Fortunately, almost everyone complied, and I did not lose too much data and was still able to complete the study! This is a cautionary, as well as humbling and amusing, tale – and highlights the need to examine how research bias can shape a study, even if a study is about challenging assumptions in the first place!

Despite this error and difficult start to my study, the results were interesting (Table 6.1).[14] The most prevalent symptom for both men and women was *katakori* (shoulder stiffness), affecting over 40% of the population. Researchers have observed shoulder stiffness to be associated with menopausal hormones in other countries,[11] so there is cross-cultural support for this being a menopausal symptom. But I wanted to know whether it was unique to women. It was not. Additionally, there were no significant differences in VMS between men and women. Although not reaching statistical significance, sudden perspiration (the only vasomotor symptom that was in the top five for frequency or severity) was higher in women than men, yet night sweats were higher in men than women. Despite sample size limitations, several symptoms did show a statistically significant difference between women and men. But all (stress, irritability, and nervous tension) were higher in men than in women! *Iraira* (onomatopoeia for irritability) is commonly included on menopausal symptom checklists in Japan, and increased *iraira* is often cited as evidence of *kōnenki* by Japanese women. Yet, it was higher in males. This raises the question of how many of the midlife symptoms that we label menopausal (and assume are related to hormonal changes of menopause) are actually uniquely or even characteristically menopausal, or whether they are simply related to aging (and even potentially associated with stressors that may affect men more).

Fieldnotes

Katakori *(shoulder stiffness) is so ubiquitous in Japan that it is included on all* kōnenki *symptom checklists and is usually the most frequently reported symptom. When discussing* kōnenki *and symptoms, many women first mention* katakori, *often cited as evidence that they are in* kōnenki. *But is it a symptom of changing hormones associated with menopause, or simply part of aging, or something in the lifestyle?*

Some people suggest that katakori *could be prevalent in Japan because of the custom of sleeping on hard futons. Others suggest it could be more common among people who commute on trains and fall asleep, with their heads bobbing, thereby putting strain on their necks and shoulders. I feel like I may have developed* katakori *myself while I was in Japan – but I am not clear about whether something in my Japanese lifestyle triggered it, or if I became more aware of the tightness of my shoulders because of the cultural attention paid to that area of the body, or because I was simply aging. As I designed my study, I was a little skeptical that* katakori *was a menopausal symptom, and my exploratory study of men and women suggest that men have it at similar rates, so perhaps it is simply a symptom of life, maybe particularly life in Japan?*

TABLE 6.1 Top five symptoms by frequency for women and men as well as those with significant differences, presented in decreasing order by prevalence in women

		Women (n = 32)	*Men (n = 22)*	*Significance*
Symptom (Japanese)	*Symptom (English translation)*	%	%	
Katakori	Shoulder stiffness	43.8	40.9	
Youtsū	Lower back pain	28.1	36.4	
Kiokuryoku gentai	Memory loss	25	22.7	
Zutsū	Headache	18.8	4.5	
Sutoresu	Stress	12.5	36.4	*
Kyū na hakkan	Sudden perspiration	12.5	0	t
Shūchūryoku gentai	Difficulty concentrating	9.4	18.2	
Iraira	Irritability	6.3	27.3	*
Shinkei no shinchō	Nervous tension	0	13.6	*
Kentaikan	Exhaustion	0	9.1	t

Modified from Melby.[14]

Significance: * = $P < 0.05$, $t = P < 0.1$.

While larger sample sizes may have led to increased statistical significance, or changed the direction of observed difference, this study yielded an essential methodological insight. Why do studies of menopause never include men, especially as they would seem to constitute good study controls to ascertain whether symptoms are due to changes in women's reproductive functioning at menopause, or to aging? The results of this small study highlight the need for appropriate controls (e.g., men and women of broad age ranges), as well as the issues of distinguishing between correlation and causation (a challenge discussed in Part I). Otherwise, we risk assuming that all symptoms observed at midlife in women are related to menopause. I am not aware of any other menopausal study that has asked men about their experiences, not least because it is almost impossible to get funding to include men in studies of menopause (for unsurprising reasons given our beliefs and biases). Nevertheless, there is evidence that even the hallmark vasomotor symptom of menopause, hot flushes, can be experienced by men.[15] Once we are willing to consider expanding our range of variation of age beyond 45–55 and sex/gender categories beyond female, we can also expand our perspectives to other cultures, to assess the extent to which culture may influence the local biology of aging and menopause.

6.2.2 Kōnenki *as Cultural Label*

When I interviewed people about whether men could have *kōnenki*, half of respondents said men have it, but not a single man admitted to having it himself. Japanese men's use of the term *kōnenki* suggests it is used to label others who are having problems at midlife (see *kōnenki shōgai* in Part IV), but it may be unlikely to be used self-referentially. The use of the term *kōnenki* to apply to men's symptoms and experiences at midlife suggested early on to me that it would differ from the concept of menopause used in the West. Although "male menopause" does appear in Western media and in peer-reviewed literature, it functions mostly to question its existence.[16] Thus, examining people's attitudes to female vs. male *kōnenki* promised to provide myriad insights.

The potential contradictions and conflicts between Western ideas of menopause and Japanese ideas about *kōnenki* came into relief for me in an interview with a physician at a hormone replacement therapy (HRT) clinic in Tokyo. I asked the physician what caused *kōnenki*, and she attributed it to loss of estrogen, and argued for the need for HRT, clearly influenced by Western biomedical models of estrogen deficiency. When I asked her whether men could have *kōnenki*, without missing a beat, she replied yes, seemingly reflecting the traditional Eastern meaning of *kōnenki* of life stage. But when I followed with the question, "What is the treatment for *kōnenki* in men," she seemed to be faced with a clash of systems, and hesitantly responded that HRT could be used, but then backpedaled a little, noting that it was not entirely clear which hormones should be used.

This brief interview highlighted the interesting points of overlap and connection between *kōnenki* and menopause, and female and male experiences, and perhaps also the inconsistencies and tensions inherent in syncretic medical systems (those combining different cultures and ideas).

6.3 Describing *Kōnenki*

To answer the question of whether *kōnenki* is equivalent to menopause, or more related to aging and less tied to reproductive functioning, comparisons of descriptions, causes, and symptoms of both *kōnenki* and menopause are relevant.

As discussed in Part I, the Western biomedical view of menopause characterizes menopause as a time at which menstruation ceases, with hallmark symptoms, such as hot flushes and night sweats. It is usually attributed to biological causation (e.g., decreased estrogen), with stages assigned primarily based on menstrual patterns and hormones.[17] Menopause is considered by many to require treatment (e.g., HRT) to avoid uncomfortable symptoms (e.g., hot flushes, vaginal dryness, decreased libido) and/or long-term chronic conditions like osteoporosis.

In contrast, Lock described *kōnenki* as a time when the body "loses its balance" with hallmark symptoms of *katakori* (shoulder stiffness) and *zutsū* (headaches). In open-ended interviews, women associated *kōnenki* with the above-mentioned symptoms, though many women reported hearing about these symptoms, not experiencing them,[3] suggesting that they might be part of a cultural narrative rather than actual experience. This highlights the importance of cultural expectations about aging and menopause (see Part IV).

6.3.1 *Causes of* Kōnenki

Lock described how *kōnenki* was attributed to both biological *and* cultural causes such as *jiritsu shinkei shitchōshō* (imbalance in the autonomic nervous system), and *hima* (excess "free" time), or being a *sengyō shufu* (full-time housewife).[3] Note, this is not to imply that housewives have that much free time in Japan. Since eldercare often falls on women at this time of life, housewives may have plenty to do besides worry about *kōnenki*! And from my observation, they have considerably higher standards of housework than I do – so that takes more time. To the extent that *kōnenki* might overlap with empty nest syndrome, children growing up and leaving home, housewives may have more free time. But the term *parasaito shinguru* (parasite single), denoting a single person in their 20s and 30s still living with their parents, reflects the increasingly common cultural phenomenon of many adult children living with and depending on their middle-aged parents. Economic and cultural changes mean that the roles and responsibilities of housewives, and parents, are changing.

In the early 2000s, women did not mention imbalance of the autonomic nervous system but did employ the new foreign loan phrase *horumon unbaransu* (hormonal imbalance), which seems to merge *kōnenki*'s sense of imbalance and Western medicine ideas about hormones. Notably, decreased estrogen was not mentioned as a primary cause of *kōnenki*. Rather, study participants often mentioned stress about caring for elders, and concern about their children and whether they would get married and find good jobs, particularly in the ongoing recession, which began in 1991, and has continued for several decades (often referred to as the "Lost Decades"). These lost decades refer to economic stagnation, but they are also associated with the loss of economic opportunities and stability for the younger generations, which may be related to lower rates of marriage and childbirth, and lower rates of establishing independent households, thereby also leading to lost freedom for some people around the age of *kōnenki*.

Fieldnotes

A year after the devastating triple disaster of the Fukushima earthquake, tsunami, and nuclear power plant disaster (which began 11 March 2011), I returned to my field site in Fukushima prefecture to visit friends and research participants. I had heard that some people were referring to the Fukushima disaster, not as a triple disaster, but as a quadruple disaster (the fourth being the stigma of Fukushima and stress associated with decisions of many mothers and children to evacuate and seek shelter in other prefectures). In fact, the impacts of the stigma and stress on people's well-being might have been even greater on many residents who were not directly impacted by the tsunami or nuclear disaster. One participant told me, "After the disaster many of my friends from my hometown called and asked how I was doing. But they were primarily concerned with the earthquake. That was not the most stressful thing in my life, but that is all they seemed to want to hear about." Similarly, while she had experienced some kōnenki *symptoms by that point and been seen by a doctor to receive* kampō *herbs and Western medicine, there were other stressors in her life that were likely more important influences on her quality of life than hormonal changes. Again, her story cautions us against assuming overly simplistic biological causation for symptoms when experience is likely influenced by a constellation of biocultural factors.*

*It also occurred to me that there might be a fifth disaster associated with Fukushima, when outsiders assume that the earthquake and related disasters are the only important events in residents' lives, and ignore other, perhaps more insider/*emic*, issues. This seems not unlike what happened with the COVID-19 pandemic, when most people's focus was on the pandemic, even as other events (illnesses, family stress, tragedies, etc.) occurred during that time but may have been ignored. The pandemic often hid or exacerbated other problems because people had figurative COVID-myopia and were unable to see other issues. Researchers particularly can be guilty of not noticing other factors due to how they frame their research topics.*

6.3.2 Search for Symptoms of Kōnenki*, Aging, and Menopause Using Factor Analysis*

To be identified as a menopausal symptom, a symptom should have: a statistical association with underlying hormones, a temporal profile associated with a peak during the menopausal transition, and a potential mechanism related to changes that occur with menopause. In contrast, symptoms of aging would be expected to worsen with increasing age, regardless of menopausal stage. Through the statistical method of factor analysis, we can uncover patterns of

TABLE 6.2 Typical factors (symptom groupings) commonly observed in population studies of menopausal symptoms, including example symptoms found in each factor

Typical factors (symptom groupings)	*Example symptoms included in factors*
Vasomotor	Hot flushes, night sweats
Psychological A	Depression
Psychological B	Anxiety
(Psycho)somatic	Headaches, gastrointestinal symptoms

symptoms that may suggest underlying etiology of a condition and minimize research biases about correlation.

Factor analysis has been used in many populations around the world to identify symptoms that tend to appear as constellations, thus suggesting shared underlying etiology. In analyses of multiple menopausal symptom scales,[18] and studies such as SWAN,[19] factor analysis usually produces three to four symptom factors as seen in Table 6.2.

In my research in Japan, because I used a very long symptom checklist, including symptoms not hypothesized to be related to menopause, analysis of symptom frequencies on a four-point scale led to eight symptom factors.[20] Once symptom factors have been identified, these can be examined for patterns over time to differentiate between aging vs. menopause, as the former would be expected to show a linear relationship with time, while the latter might show an inverted U-shaped curve, peaking during the menopausal transition and then decreasing.

VMS strongly associate with, and usually increase through, the menopausal transition, and then eventually decrease at some point post-menopause. Thus, we would expect symptoms associated with underlying hormonal changes of menopause to show differences between pre- and perimenopausal women, and possibly also postmenopausal women. Several symptom factors showed significant differences between menopausal and *kōnenki* groups (Table 6.3), but interestingly these were not the same symptom factors.

While the sexual-vasomotor (SV) symptom factor showed significant differences between pre- and perimenopausal status groups (as would be expected for symptoms caused by underlying hormonal changes of menopause), it did not show significant differences between pre and peri *kōnenki*. This provides evidence that menopause and *kōnenki* are not the same! While the autonomic nervous system symptom factor showed no significant differences by menopausal status, it did show significant differences by *kōnenki* status, reinforcing the idea that *kōnenki* labels reflect changes in autonomic nervous system-related symptoms, and suggesting that *kōnenki* may be more than (or different from) menopause and that the autonomic nervous system is important for understanding and diagnosing *kōnenki*.

TABLE 6.3 Symptom factors showing significant differences in regression scores by *kōnenki* and menopausal status based on frequency data. With top four symptoms provided as examples

Symptom factor	*Top four symptoms (by factor loadings)*	*Significant differences by three* kōnenki *status groups*	*Significant differences by four menopausal status groups*
Sexual-Vasomotor	• Vaginal dryness • Painful intercourse • *Nobose* (hot flush) • *Kyū na nekkan* (hot flush)	Pre-Post*	Pre-Peri II***
Psychological-Anxiety	• Feeling of worthlessness • *Hotto furasshu* (hot flash) • Palpitations • Worry about getting old		Pre-Peri I*
Psychological-Depression	• Melancholy • Depression • Insomnia • Difficulty falling asleep	3 groups*	
Somatic 1	• Shortness of breath • Pain in the ischium (lower/back hip) • Coruscation (flash of light) • Ringing in the ears	3 groups**	Pre-Peri I*
Somatic 2-Autonomic Nervous	• Nervous tension • Oppressive feeling • Rush of blood to the head • Dizziness	3 groups*	
Somatic 3	• Vaginal discharge • Constipation • Exhaustion • Shoulder stiffness		3 groups*

Modified from Melby.[20]

Significant differences in factor regression scores by *kōnenki* or menopausal status (* $P < 0.05$, ** $P < 0.01$, *** $P < 0.001$). Tests for significance included: Brown-Forsythe test for differences among three *kōnenki* groups (pre-, peri-, and post-*kōnenki*); Median test for differences between 3 and 4 groups of menopausal status (Pre, Peri I & II, Post); and Mann-Whitney two-tailed test for differences between two groups.

6.4 Are Japanese Hot Flush Symptoms All Clearly Vasomotor Symptoms (VMS)?

As previously explained, I use the term "hot flush" (the more common British word) as opposed to "hot flash" (more common in the United States) to gloss three Japanese terms for hot flush (*nobose*, *hoteri*, and *kyu na nekkan*). I chose to do this because there is another word in Japanese, the foreign loan word *hotto furasshu* that derives from the English word "hot flash," but it seems to be quite different. Interestingly, in factor analysis, it loaded not onto to the sexual-vasomotor factor, but rather onto the psychological-anxiety factor, meaning that women reporting *hotto furasshu* tend to report symptoms of anxiety.

When I began my fieldwork, I spoke with a woman who told me she had experienced a *hotto furasshu* that lasted a year, suggesting that the idea of a "flash" may have been lost in translation! But since this term did not sort with other hot flush terms, and instead sorted with psych-anxiety factors, it seems that this term may be used to describe something other than a typical hot flush, perhaps suggesting some degree of somatization or medicalization and related anxiety about midlife and aging (see Part IV).

Had I gone to Japan and asked women only whether they had experienced a *hotto furasshu* in the previous two weeks, my study suggests that I would have found that very few Japanese women reported experiencing hot flashes, because only 3% of women reported *hotto furasshu* in my study.

The words *nobose* and *hoteri* likely convey a similar experience to the Western hot flush; thus, if we are trying to compare rates of hot flushes, we need to make sure we use, not similar linguistic terms, but words that actually capture similar phenomena. Had I not included words like *nobose* and *hoteri* in the symptom checklist, I could not have captured comparable experiences. However, the difference in rates and meaning of those words also raises questions about variation in hot flushes, and more subtle biology and experience than we in the West have even considered, especially since these are not differentiated in our language. In research, what you "see" depends on what you ask about, and the vocabulary you must work with!

Consistent with other studies of menopause around the world, my factor analysis resulted in a single sexual-vasomotor symptom factor containing hot flushes, but interestingly this factor did not include night sweats. Most surprisingly, my analysis brought to light a potentially novel vasomotor symptom, *hieshō* (chilliness), which loaded onto the sexual-vasomotor factor. While the idea of being frigid has a sexual connotation in the English language, which may be associated with loss of libido at menopause (see Part I), in Japanese, *hieshō* conveys a more straight-forward issue with thermoregulation. This conglomeration of symptoms on opposite ends of the thermoregulatory spectrum (ranging from hot flushes to chilliness) all loading onto the same sexual-vasomotor factor suggests they share a similar underlying etiology.

Perhaps thermoregulatory dysregulation caused by hormonal changes? These results further suggest that chilliness may be a vasomotor symptom, one that may have been overlooked in other populations due to language limitations and/or differences in local biology. This exemplifies the insights to be gained from applying a biocultural approach: demonstrating how proximate culture in the form of language can provide clues to proximate biology!

Things are rarely as simple as we would like them to be. While chilliness may be a novel (to the West) vasomotor symptom, it may not be a symptom of menopause exclusively. Unlike *nobose* and *hoteri*, rates of which went down in postmenopausal women after peaking in perimenopause, chilliness remained high, suggesting that aging processes may play a role, perhaps accelerated during perimenopause. Other studies of chilliness with larger sample sizes have shown similar patterns among women experiencing chilliness, but interestingly, chilliness in males appeared to peak in the 50s and 60s and then decline, suggesting menopause-like symptom patterns in males, but not females.[21] Clearly, further exploration of thermoregulatory symptoms and their association with age, hormones, and biological sex and cultural gender is needed.

6.5 Local Biology: Culture in the Form of Diet

Ideally, if we wanted to examine differences in symptoms related to diet, we would identify populations or individuals that differed *only* in terms of diet. But this is rarely possible. Even though Japan is often thought to be culturally homogeneous, there are geographic, cultural, and even ethnic differences within the country.[22] While Fukushima and Kyoto were chosen as field sites because of dietary differences, other aspects of culture also vary, and may influence symptom reporting and experience (and local biology).

Kyoto was the country's capital for many years and is still considered to be the nation's cultural capital. In contrast, Fukushima, as part of the Tohoku region, exists more on the cultural periphery. In some ways, Kyoto and Fukushima could be considered analogous to New York and the American South in the US. The Tohoku region has dialects that are given subtitles on Japanese TV due to the extent to which they depart from standard Japanese (spoken in Tokyo). While Kyoto also has a distinctive dialect, it is one that is well known enough not to be subtitled.

Since cultural variation between regions may also influence reported differences in experience, ethnographic data are important to provide context to the hormone samples, dietary records, and daily symptom checklists that we collected. Taken together, these data allow us to examine the concept of local biology and explore the pathways through which culture becomes embodied.

6.5.1 *Japanese Variation in Soy Intake*

In designing my study, I hypothesized that soy phytoestrogen intake and/or exposure would be inversely correlated with symptom prevalence (i.e., the higher the soy intake or exposure, the lower the symptoms). To test this

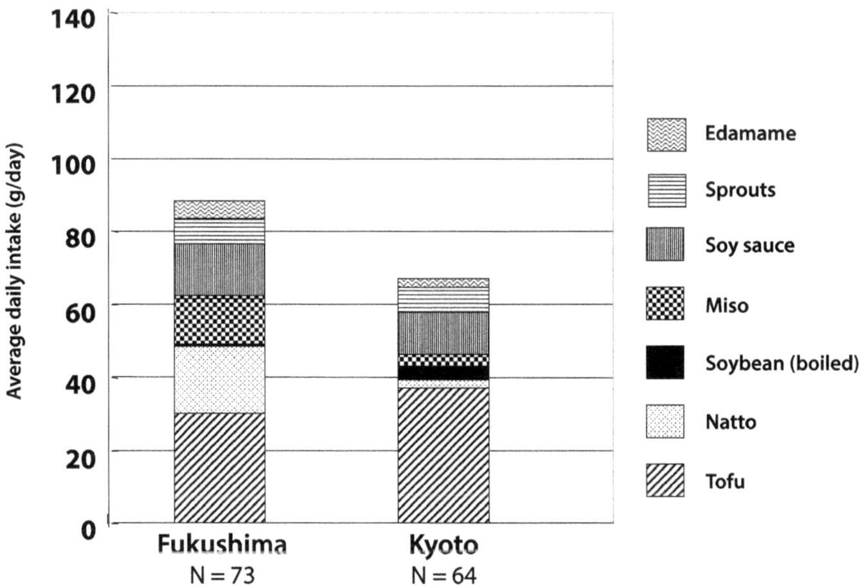

FIGURE 6.1 Average amounts of soy foods (g/day) consumed in Fukushima and Kyoto study populations in KF-WMHS, based on Food Frequency Questionnaire.

hypothesis, variation in soy intake is required. This is why I conducted research in two regions in Japan with low and high soy intake: Kyoto prefecture in the Kansai region with 12.3% less, and Fukushima prefecture in the Tohoku region with 22.44% more per capita soy consumption than the national average in the late 1990s.[23] The regions also vary in the proportion of different types of soy foods, including fermented (*miso*, *nattō*) vs. unfermented soy ingested (Figure 6.1). As discussed in Part I, the microbiome plays a role in fermenting soy both outside the body (in *miso*, *nattō*) and inside the body. And gut microbes metabolize soy phytoestrogens in the body, producing equol from daidzein (if the necessary microbes are present) (see Chapter 4).

The diet in Fukushima is known for having *koi aji* (salty, strong flavorings) and contains more fermented soy in the form of *nattō* (fermented soybeans) and *miso* (fermented soybean paste) compared to Kyoto. Kyoto cuisine is characterized by *usui aji* (delicate flavorings) and the diet contains more tofu, and related foods like *yuba* (the "skin" of coagulated proteins that forms on the surface of heated soy milk, which I realize does not sound very delicious, but I assure you that with a little bit of soy sauce it is to die for).

Fieldnotes

I began my fieldwork in Kyoto and was affiliated with the Kyoto Prefectural University of Medicine, where I was generously given a desk and space to store all my research materials – and space for my research assistants as well. I began work early, and my Kyoto host mother provided breakfast – often traditional tofu *and* miso *soup with vegetables at 4:30 or 5 every morning. She also introduced me to* yuba *– the skin formed on boiling soy milk – with a texture almost like melted mozzarella cheese. This could be seen as a waste product in the process of making* tofu, *but is one of the most delicious forms of soy I have ever eaten.*

During the day, I was usually out doing interviews, and then returned to the office to organize materials, log samples, etc. and sometimes worked until quite late. If I got hungry at night, there were not many restaurants in the immediate vicinity, so I started storing packets of nattō *(fermented soybeans) in the freezer. When I got hungry, I would microwave it for a minute in the microwave, which released a very distinctive* nattō*-smell throughout the entire office and lab. At night most people had gone home – but one colleague often worked late. He would call to me from the other room and say, "Is that* nattō *...?" I would apologize for the smell, which he was clearly not fond of, or at least enjoyed teasing me about, particularly since* nattō *is generally not well-liked in the Kyoto region – and certainly not as a snack. His late-night snacks of choice were much more American: chocolate or potato chips. The fact that I preferred to snack on* nattō *was quite amusing to my Kyoto colleagues. Like potato chips,* nattō *is also a bit salty, but is high in protein and vitamins, and is almost as "instant" as opening a bag of potato chips!*

In contrast, at my second field site in Fukushima, nattō *was quite common for breakfast, as well as other meals and could be prepared in many ways – tempura, in soup, with rice and rolled in seaweed (*nori-maki*). There, my* nattō*-eating habits surprised nobody, and led to fewer comments. Although one friend used to introduce me to others by saying "Melissa looks like a gaijin (foreigner) on the outside, but on the inside she is Japanese," mostly referring to my culinary preferences, but perhaps also conveying a broader meaning.*

6.5.2 Challenges of Studying Diet

Some of the challenges of studying diet include: accuracy of self-reporting, challenges associated with measurement, time scale (daily/weekly/seasonally/annually), measurements of nutrients or metabolites in blood (and their relationship to timing of ingestion), and the issue of correlation vs. causation. For example, while Japanese have relatively good health and among the longest life expectancy compared to other populations, this has often been attributed to aspects of the diet such as soy, seafood, and vegetables. But compared to other diets, what do those foods "replace"? Could it be that Japanese health is not due primarily to the presence of "good foods" such as soy and seafood, but the absence of "bad foods" such as red meat, or perhaps the combination of the two?

Fieldnotes

For the dietary records, I asked women to record everything they ate for 24 hours before taking blood samples. They did this one day every week for six months. I am not sure I could have followed such a protocol myself. It was very tedious, and particularly challenging for women who were cooking meals for their families, to estimate how much they personally ate at each meal. At the end of the study, I asked participants to provide feedback on the research and their participation. One woman confessed to me that the dietary records were particularly challenging (time-consuming and tedious) and that over the six-month period she found herself choosing to eat foods that were particularly easy to record on the days she was recording dietary data. For example, one banana was easy to write. Fortunately, nattō *and tofu also come in easy to record sizes, and thus soy consumption was not likely cut from their diets on the dietary record days. But clearly in a longitudinal study, one should be aware that the study procedures may be an intervention in themselves and change behavior, particularly with diet where people might attempt to adhere to a "better" diet on days they are recording.*

Proximate culture in the form of diet and other environmental exposures may lead to differences in gut microbiome, and in turn differences in the resulting metabolites produced upon ingestion of soy, with varying impacts on symptoms. Dietary records do not account for metabolism of soy phytoestrogens by gut microbes, and thus blood samples are required to measure metabolites. While the primary phytoestrogens in soy are genistein and daidzein, the latter can be converted to a compound called equol, which appears to be more bioavailable and binds to estrogen receptors with a greater affinity.[24] The ability of certain gut microbes to produce equol[25] may partly account for the population differences in the effects of soy seen in Japan and the United States (up to 60% of people are equol-producers in Japan vs. only 20% among Americans)[24] and for the conflicting results in many clinical intervention studies conducted in different populations.[26,27] In a population such as Japan where most people eating soy may be exposed to equol, soy may show significant benefits. But in populations such as the United States where only 20% of people (or their gut microbes actually) can produce equol from daidzein, soy may show less significant benefits. Being an equol-producer might give a competitive advantage in terms of receiving benefits from soy.[28] Because of the role of the microbiome, what goes in (the mouth) may not be what comes out (in the blood), and thus future studies of diet may need to move beyond dietary intake records to increasingly measure dietary metabolites.

We need methods to measure the "local biology" influencing health. Since effects are likely to be influenced by the gut microbes present, it is important to assign equol-producer status based on blood measures. Some researchers give participants a soy challenge to assess whether people have the capability of producing equol when they ingest soy (this is similar to a glucose challenge to assess for diabetes). But this status may only be important if soy is consumed regularly. With increased awareness of the role of the microbiome and metabolites, research approaches need to incorporate a more nuanced understanding of how diet can affect people's health and symptoms, not just through the content of the food, but through bacterial mediators residing in people's guts, which affect what cells and organs actually "see" after eating and the associated local biology health effects.

Interestingly, much biomedical research relies on studies in rats. Those rats consume commercial lab rat chow, an industrialized processed food that often contains soy protein. Compared to humans, rats appear to be equol-producing machines because their microbiome leads to higher conversion of daidzein to equol. Thus lab chow soy phytoestrogens may influence physiology, neurobiology, and behavior,[29] skewing experimental results in unknown ways.[30] Given how reliant biomedical science is on studies of rats to better understand physiology and behavior of humans, it is interesting to reflect on how much our understandings may have been influenced by dietary hormones deriving from soy in the lab diet (obviously not something rats tend to eat in the wild, unless they are eating from a garbage dump outside a tofu-making factory). Rats consuming standard lab rat chow containing high quantities of phytoestrogens may only be relevant models for some populations, like Japanese, who ingest higher quantities of phytoestrogens.

In North America, women often start eating soy or taking soy supplements when they notice menopausal symptoms like hot flushes. However, it is possible that they will not see such remarkable effects as Japanese who consume soy all their lives, and have microbes that metabolize phytoestrogens into more bioactive molecules. We need to consider not just what people eat in the day or so before we measure their hormones or symptoms (i.e., acute effects of consumption of soy phytoestrogens), but also chronic or developmental effects in key time frames, and microbial metabolites such as equol that may have different effects than the ingested compounds. To this day, most studies cannot control for effects of recent vs. lifelong soy consumption on the health outcomes of the participants at midlife unless they collect retrospective data, since prospective data collection would take decades. Given the widespread effects of estrogen throughout the body (due to the presence of estrogen receptors throughout the body), lifelong soy consumption might have impacts not only on menopausal symptoms but also on many other aspects of physiology and development.

Timing of soy intake and exposure may be critical for teasing out different health outcomes. For example, when consumed in adolescence, soy may promote differentiation of breast tissue, decreasing risk of cancer.[31] Most research suggests that soy phytoestrogens do not increase risks of breast cancer in humans like estrogen does,[28] perhaps because soy phytoestrogens are more accurately considered selective estrogen receptor modulators.[32]

The experience of aging in Japan is influenced by a syncretic blend of East Asian and Western biomedical understandings of the body, health, and aging. Environmental factors (ranging from dietary soy to traditional living on tatami mats and bathing sitting down) may influence the development of aging-associated conditions such as osteoporosis and fracture rates. Although *kōnenki* is often translated as menopause, beliefs about who has it and when, what causes it (hormone imbalance as opposed to decreased estrogen), and its associations with symptoms, suggest that it does not capture the same experience as the word menopause. The novel (to the outsider/*etic* perspective) symptom of chilliness appears to be an important part of *kōnenki* and both challenges and expands the biomedical notion of VMS. Culture in the form of diet, particularly soy and its constituent phytoestrogens, show variation in Japan that may contribute to the local biology of *kōnenki*.

References

1. Pimpton G. 1990. Maya Angelou, The Art of Fiction No. 119. *Paris Rev.* Fall 1990(116):144–167.
2. Lock MM. 1984. East Asian medicine and health care for the Japanese elderly. *Pac Aff.* 57(1):65–73. https://doi.org/10.2307/2758386
3. Lock M. 1993. *Encounters with Aging: Mythologies of Menopause in Japan and North America.* Berkeley, CA: University of California Press.
4. Littler J. 2019. *The Art of Perseverance: How Gaman Defined Japan.* Available at: https://www.bbc.com/worklife/article/20190319-the-art-of-perseverance-how-gaman-defined-japan. Accessed 18 Jan, 2025.
5. Maeda Y, Takahashi K et al. 2019. Factors affecting the provision of analgesia during childbirth, Japan. *Bull World Health Organ.* 97(9):631–636. https://doi.org/10.2471/blt.19.230128
6. Butwick AJ, Bentley J et al. 2018. United States state-level variation in the use of neuraxial analgesia during labor for pregnant women. *JAMA Netw Open.* 1(8):e186567. https://doi.org/10.1001/jamanetworkopen.2018.6567
7. Melby MK, Zent E et al. 2025. "The Littlest Creatures that Live Inside Us": Public understandings influencing microbiome-related behaviors. *Social Science & Medicine.* 376:117864. https://doi.org/10.1016/j.socscimed.2025.117864
8. Kleinman AM. 1977. Depression, somatization and the "new cross-cultural psychiatry". *Soc Sci Med (*1967). 11(1):3–9. https://doi.org/10.1016/0037-7856(77)90138-X

9. Kirmayer LJ, Young A. 1998. Culture and somatization: Clinical, epidemiological, and ethnographic perspectives. *Psychosom Med.* 60(4):420–430. https://doi.org/10.1097/00006842-199807000-00006
10. Lock M, Kaufert P. 2001. Menopause, local biologies, and cultures of aging. *Am J Hum Biol.* 13(4):494–504. https://doi.org/10.1002/ajhb.1081
11. Dennerstein L, Lehert P et al. 2007. A symptomatic approach to understanding women's health experiences: A cross-cultural comparison of women aged 20 to 70 years. *Menopause.* 14(4):688–696. https://doi.org/10.1097/gme.0b013e31802dabf0
12. Alder E. 1998. The Blatt-Kupperman menopausal index: A critique. *Maturitas.* 29(1):19–24. https://doi.org/10.1016/S0378-5122(98)00024-3
13. Melby MK, Lock M, Kaufert P. 2005. Culture and symptom reporting at menopause. *Hum Reprod Update.* 11(5):495–512. https://doi.org/10.1093/humupd/dmi018
14. Melby MK. 2006. Climacteric symptoms among Japanese women and men: Comparison of four symptom checklists. *Climacteric.* 9:298–304. https://doi.org/10.1080/13697130600868653
15. Jones JM, Kohli M, Loprinzi CL. 2012. Androgen deprivation therapy-associated vasomotor symptoms. *Asian J Androl.* 14(2):193–197. https://doi.org/10.1038/aja.2011.101
16. Pines A. 2011. Male menopause: Is it a real clinical syndrome? *Climacteric.* 14(1):15–17. https://doi.org/10.3109/13697137.2010.507442
17. Harlow SD, Gass M et al. 2012. Executive summary of the Stages of Reproductive Aging Workshop + 10: Addressing the unfinished agenda of staging reproductive aging. *Menopause.* 19(4):387–395. https://doi.org/10.1097/gme.0b013e31824d8f40
18. Greene JG. 1998. Constructing a standard climacteric scale. *Maturitas.* 29(1):25–31. https://doi.org/10.1016/s0378-5122(98)00025-5
19. Avis NE, Stellato R et al. 2001. Is there a menopausal syndrome? Menopausal status and symptoms across racial/ethnic groups. *Soc Sci Med.* 52(3):345–356. https://doi.org/10.1016/s0277-9536(00)00147-7
20. Melby MK. 2005. Factor analysis of climacteric symptoms in Japan. *Maturitas.* 52(3–4):205–222. https://doi.org/10.1016/j.maturitas.2005.02.002
21. Yoshino T, Katayama K et al. 2013. Statistical analysis of hie (cold sensation) and hiesho (cold disorder) in kampo clinic. *Evid Based Complement Alternat Med.* 2013:398458. https://doi.org/10.1155/2013/398458
22. Narzary DC. 2004. The myths of Japanese 'homogeneity'. *China Rep.* 40(3):311–319. https://doi.org/10.1177/000944550404000308
23. Japanese Ministry of Health. *Kokumin eiyoo no genjoo - Heisei rokunen kokumin eiyoo choosa seiseki. (The Present State of Nutrition in Japan: The Report on the 1994 National Nutrition Survey).* Tokyo: Japanese Ministry of Health.
24. Setchell KDR, Brown NM, Lydeking-Olsen E. 2002. The clinical importance of the metabolite equol—A clue to the effectiveness of soy and its isoflavones. *J Nutr.* 132(12):3577–3584. https://doi.org/10.1093/jn/132.12.3577
25. Rafii F. 2015. The role of colonic bacteria in the metabolism of the natural isoflavone daidzin to equol. *Metabolites.* 5(1):56–73. https://doi.org/10.3390/metabo5010056
26. Lampe JW. 2009. Is equol the key to the efficacy of soy foods? *Am J Clin Nutr.* 89(5):1664s–1667s. https://doi.org/10.3945/ajcn.2009.26736T

27. Newton KM, Reed SD et al. 2015. A cross-sectional study of equol producer status and self-reported vasomotor symptoms. *Menopause.* 22(5):489–495. https://doi.org/10.1097/GME.0000000000000363
28. Messina M, Duncan A et al. 2022. The health effects of soy: A reference guide for health professionals. *Front Nutr.* 9:1–33. https://doi.org/10.3389/fnut.2022.970364
29. Finney CA, Proschogo NW et al. 2020. The effect of standard laboratory diets on estrogen signaling and spatial memory in male and female rats. *Physiol Behav.* 215:112787. https://doi.org/10.1016/j.physbeh.2019.112787
30. Brown NM, Setchell KDR. 2001. Animal models impacted by phytoestrogens in commercial chow: Implications for pathways influenced by hormones. *Lab Invest.* 81(5):735–747. https://doi.org/10.1038/labinvest.3780282
31. Messina M, Hilakivi-Clarke L. 2009. Early intake appears to be the key to the proposed protective effects of soy intake against breast cancer. *Nutr Cancer.* 61(6):792–798. https://doi.org/10.1080/01635580903285015
32. Oseni T, Patel R et al. 2008. Selective estrogen receptor modulators and phytoestrogens. *Planta Med.* 74(13):1656–1665. https://doi.org/10.1055/s-0028-1088304

7

PROXIMATE CULTURE OF MENOPAUSE

In Quadrant II of the Proximate-Ultimate Biocultural Framework (Figure 5.1), we ask: What is the proximate culture of menopause, particularly in the context of Japan? How does culture in various forms influence the lived experience of menopause?

Language not only provides a window into culture, but it is also a primary way culture is transmitted. In this chapter, we explore the language of *kōnenki* and menopause, mapping different stages of each to assess the degree of overlap and points of difference. We then move on to the language of hot flushes and ask what do the numerous Japanese hot flush terms mean for the lived experience of menopause? Also, what do they suggest about what might be missing from Western understandings? Could our language for hot flushes, or lack thereof, lead us to miss important experiences and even biology? How might our single word for hot flush reflect more limited experience, or obscure more diverse biology?

The experience of menopause is of course influenced by factors we might consider strictly biological (e.g., ovarian reserve, hormones), but also appears to be influenced by cultural factors that have biological effects. To better understand the cultural factors influencing variation in menopausal experience, a cross-cultural perspective can highlight important gaps in our knowledge and provide more nuanced views than a focus on a single culture. For example, Japanese experiences and views of *kōnenki* blur lines between Western ideas of aging and menopause.

Language can be limiting, misleading, or illuminating. Clinically, the word menopause denotes the end of menstruation (defined as one year after what is assumed to be the last menses). But for decades, most North American

DOI: 10.4324/9781003644521-9

women have viewed menopause not as a single event, but as a process,[1] encompassing symptoms and experiences during the menopausal transition.

In Japan, the relationship of *kōnenki* to *heikei* (end of menses) appears unclear as some women report that they believe they can avoid *kōnenki* altogether (regardless of their menstrual patterns). To a Western biomedical ear, this statement seems nonsensical because all women will experience menopause (in the sense of *heikei*, or final menses) if they live long enough. However, this is where language is critical. Through in-depth interviews about the term *kōnenki*, I realized that some women were using the term *kōnenki* as a shorthand for *kōnenki shōgai* (menopausal disorders). Thus, with menopause in Japan, it is very clear that insider-defined *kōnenki* is not the same thing as outsider-defined Western biomedical menopause. There also seems to be considerable individual variation in how these terms are used in Japan.

It appears that both English-speaking and Japanese-speaking women have simplified their terminology choosing a single term to denote both the event of menses cessation as well as the transition surrounding the event, and possibly also symptoms associated with the transition. English-speaking populations have taken the word denoting the narrow event (menopause) and expanded its use to the perimenopausal transition and to encompass meanings previously denoted by climacteric. Japanese have gone the other direction, taking the broader term *kōnenki* (closer to climacteric in meaning) and essentially made the term *heikei* obsolete, using *kōnenki* to connote both the end of menses and the life transition surrounding it, or even the symptoms commonly associated with the transition.

Thus, while *kōnenki* is the Japanese word that is often used to translate the English word menopause, closer inspection reveals that these terms characterize different, although overlapping, phenomena. The experience of *kōnenki* in Japan is defined more broadly than Western menopause in terms of age, possibly including even gender (see Kanazawa study in Chapter 6), cultural and psychological factors, as well as physical symptoms.

7.1 Mapping *Kōnenki* to Menopause

There are many approaches to assessing the overlap between *kōnenki* and menopause. In Chapter 6, we discussed how factor analysis of symptoms allows us to assess overlap in symptom groupings. Here, we map individual women's categories, comparing the insider/*emic* self-assessed *kōnenki* status with the more outsider/*etic* "objectively" assessed menopausal status based on menstrual patterns (Figure 7.1) Consistent with the idea that menopause and *kōnenki* are related (but also with the idea that both are correlated with aging), the distribution of pre- to post-*kōnenki* self-assignment does shift from pre- to post-menopause. A greater percentage of premenopausal

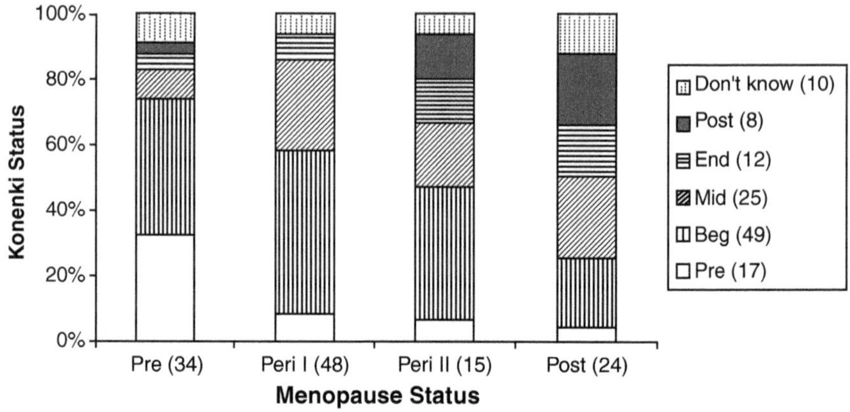

FIGURE 7.1 Mapping of *kōnenki* status (% self-assigned to pre, beg, mid, end, post, and "don't know" categories) onto menopausal status assigned based on menstrual patterns ($N = 121$) (reproduced with permission).[2]

women considered themselves to be in pre- and beginning *kōnenki* phases compared to peri- and postmenopausal women. But for each menopausal category, there were women who assigned themselves to every *kōnenki* category ranging from pre to post. Most striking is that while all categories of *kōnenki* status can be observed in all menopausal status categories, the postmenopausal women seem to have all the "colors" of the *kōnenki* rainbow (if this figure were in color) in almost equal proportion.

It is important to note that menopausal status was assigned at entry to the study protocol, and *kōnenki* status was self-assigned at the end of six months of participation (when I "came clean" and told people I was really interested in learning more about *kōnenki*). Thus, some premenopausal women (who were having regular menstrual cycles and had no menstrual signs of menopause) could report being in mid-*kōnenki* six months later if menopause and *kōnenki* were the same thing. Premenopausal women reporting being at the end or even post-*kōnenki* six months later was less expected. The perimenopausal women who reported six months later that they were still pre-*kōnenki* was also unexpected. But the most surprising were the postmenopausal women, who had not menstruated for over a year at entry to the study), but who reported being pre- or beginning *kōnenki*. This clearly shows that there is not a one-to-one mapping of menopause (as we understand it in the Western biomedical sense) and *kōnenki*.

These observations beg the question of how people diagnose themselves with *kōnenki*. The primary criteria for self-assignment of *kōnenki* were symptoms (40.8%) and menstrual period characteristics (33.3%) (total of >70% of responses). Physical condition also was a criterion for 11.6% of the study sample. Since Western biomedical menopausal status is often assigned based on self-reported menstrual period patterns, this shows a clear overlap. This also highlights how biomedical categories are not truly objective, especially when based on women's subjective reporting of their menstrual patterns. But since symptoms were the primary criterion for the assignment of *kōnenki*, it is important to explore what *kōnenki* symptoms were, and to explore how they might differ from hallmark symptoms of Western biomedical menopause. If symptoms are not unique to *kōnenki*, but perhaps more related to aging, this has implications for whether and how people choose to treat them.

Having established that the language of and criteria for assignment of menopause and *kōnenki* are unclear, or at least fuzzy at the boundaries, we next turn to the language of hot flushes.

7.2 Language of Hot Flushes

Fieldnotes

In the final interview, one participant told me that she had finally experienced a hot flush. She recalled the situation in detail. Standing in the kitchen, she suddenly began to feel hot. "Aha," she thought, "this is it, my first hot flush." She called to her husband in the adjoining room and asked him if the house was hot. He said no. After six months of completing daily symptom checklists about midlife symptoms, she was thrilled to finally have something to report. She then excitedly said to her husband, "I think this must be kōnenki*– I am having a hot flush!" A few seconds later, she noticed the kerosene stove (used by many households in northern Japan, where central heating is rare) and remembered that she had recently turned it up to help dry the laundry she had hanging up in the kitchen. No wonder she was hot – she was standing over a kerosene stove, turned up to high! Disappointed that she may not yet have arrived at* kōnenki's *door, but amused by her own excitement, she informed me that she may have completed the six-month study without having entered* kōnenki.

Although I did not express my specific interest in menopause until late in the final interview to avoid women looking for "menopausal symptoms," clearly this was not failproof. Apparently, this participant had been on the "lookout" for this sine qua non symptom of menopause that she had heard so much about in the media and had surmised that I might be interested in this as part of my study on Midlife Women's Health. Similarly, researchers often go into the field looking for something (data to support their hypothesis), and they run the risk of mistakenly finding it, especially if they are not open to alternative possibilities that may be closer to the truth.

This interaction from the field highlights the extent to which participants' expectations and understandings of what researchers are looking for can influence the results.

Popularizations of Margaret Lock's work often state that there is "no word for hot flushes in Japan." This provides a good example of less-than-rigorous writing, as it is partially true. There is no "single" word, but actually, there are several terms used to describe hot flushes. But this raises the question of which word(s) to use. Depending on the term(s) used, how can we know if we are comparing and measuring the same phenomenon when conducting cross-cultural research? If there are cross-cultural differences in the experience of symptoms such as hot flushes (frequency, severity, etc.), we want to know why. Is it because of genetic differences, environmental differences such as diet or ambient temperature, or cultural influences on perception of symptoms and how much people recognize or report them? But first, we need to establish if there are truly any differences.

For many issues of interest to public health researchers and practitioners, comparative data, for example on symptoms, are useful. But you can't just use a translated symptom checklist and conclude that the numbers you get somehow capture all the important experiences, and at worst, they may be completely misleading. There is no clear solution to the myriad challenges of how culture influences symptom reporting at menopause.[3]

I thought hot flushes would be the most straight-forward symptom to measure. I first obtained an overview of symptom rates by using an extensive two-week recall checklist (modified from Lock's checklist) given to women to fill out at the beginning of their six-month participation in the study. Then I examined how rates of select symptoms varied over the study period using an abridged daily symptom checklist. But I struggled with which words to use for hot flush, since there is no clear translation in Japanese. Margaret Lock had used three Japanese terms (*hoteri*, *nobose*, and *kyū na nekkan*) to gloss the English word hot flush. I included those three terms, and added the more recent term *hotto furasshu*, a Japanese foreign loan phrase derived from the North American English word hot flash.

Once terminology was out of the way (or so I thought), I had to decide how I wanted women to report them – in terms of frequency, severity, bothersomeness, etc. I decided on frequency and severity. Four frequency (F) responses (0–3) ranged from not at all, to one to two times per week, three to five times per week, and almost daily, and three severity responses (1–3) ranged from mild to moderate to severe. A frequency-severity (FS) score from 0 to 9 was computed by multiplying frequency response by severity response. By including severity, the goal was to capture more variation, but a tendency among Japanese to believe or report that they are like everyone else might have influenced most women to report moderate severity. If I had asked "degree to which you are bothered" by symptoms, the cultural ideal of "*gaman*" (endurance, see Chapter 6), may have biased people toward reporting none. This phenomenon may extend beyond menopausal symptoms, as lower symptom reporting among East Asians compared to North Americans has been reported even for conditions such as heart failure.[4]

7.2.1 Tatemae *(Façade) vs.* Honne *(True Feelings)*

In clinical medical encounters, often between female patients and male doctors, the power and status dichotomy may make it difficult for women to feel comfortable telling doctors everything about their symptoms. The distinction between *honne* (one's true feelings) and *tatemae* (façade of behavior and opinions one shows in public) is important to contextualize symptom severity and how women report bothersomeness (at least to outsiders, like researchers). This distinction also highlights why ethnography is important, and how aspects of collectivist society (e.g., desire to fit in and appear similar to others in Japan vs. US "squeaky wheel" and individualism) might influence perceptions, experience, and research results.

In terms of research design, it is also relevant that Kyoto is famous for *tatemae*, which is the public façade people show to outsiders. In contrast, Fukushima, which is part of the Tohoku region, is characterized by a greater degree of *honne*. Thus, my two field sites differed not only in soy consumption but also in degrees of *tatemae* and *honne*.

When I was going through research ethics approval in Kyoto, I was told by male physician colleagues that they thought it unlikely that women would answer such probing questions and provide such detailed data as I was requesting. By going to women's homes, workplaces, or other sites of their choosing, I changed the dynamic that would have ensued had I made them come to the university or clinic for their interviews. This is one benefit of participant observation (field research method used widely in anthropology and sociology that involves studying a group's life, behavior, values, etc. by sharing in its activities). By participating in their daily lives, sharing meals and hikes with them, I became part of their lives (albeit briefly) and might have been able to elicit more *honne* rather than *tatemae* in their responses. What my male doctor colleagues did not realize was that my gender, age, nationality, and my sometimes seemingly over-the-top efforts to interview women in their homes perhaps made it easier and safer for participants to disclose to me details about their symptoms and lives.

These included topics that were relatively stigmatized and taboo and often hidden from the rest of society, ranging from mental health issues and disabilities to affairs.

Fieldnotes

During my fieldwork, several participants welcomed me into their homes and their lives and became like family to me. Many were people living in Fukushima, a region of Japan known more for its honne. *In contrast, while many Kyoto residents were very friendly and generous, I later discovered that many of the participants who befriended me were not native multi-generational Kyoto residents but had moved from elsewhere. They often told me that even for Japanese, Kyoto society was hard to penetrate.*

I feel particularly fortunate for the relationships I did develop with "native" Kyoto people. For the first few months until I found my own place, I stayed in the home of a very traditional Kyoto family. My homestay mom was a fierce kyoiku *(education) mama, not like the stereotypical Asian Tiger mom, but more in a strategic, confident way that may derive in part from privilege, education, and knowledge of how to work the system. She prepared her children to be successful, while taking care of multiple elders, and doing community volunteer work. Once her child-rearing duties were over, she went back to school herself. Definitely a force to be reckoned with! Initially I felt that she maintained an element of* tatemae *or formality that was difficult to penetrate, particularly in contrast with some of my participants who were not Kyoto natives. But over time, we became quite close.*

Another participant, originally from Kyoto who had moved to Fukushima, was extremely welcoming to me. She carried an air of cultured sophistication and formality more commonly observed in the cultural capital of Kyoto than in more rural Fukushima. Perhaps because of my connection to Kyoto, or the fact that she no longer lived in Kyoto, I felt like our relationship was able to move beyond the tatemae *so typical of Kyoto.*

My efforts to hear women's stories and lives were rewarded with great details about their experiences, and incredible compliance with a time-consuming research protocol.

7.2.2 Symptom Frequency, Factor Analysis, and Sexual-Vasomotor Symptom Factor

Since I was interested in identifying which symptoms of *kōnenki* might share underlying (e.g., hormonal) etiology and show an increase during menopause,

TABLE 7.1 Two-week frequencies for symptoms in the sexual-vasomotor (SV) factor (along with several other symptoms for reference) and the factor loadings when analyses were conducted using symptom frequency (N/A = not applicable, i.e., not observed in the SV factor)[5]

Symptoms	*Two-week frequency (%) (N = 140)*	*Frequency factor loadings on SV factor*
Shoulder stiffness (*katakori*)	62.1	N/A
Headache	39.3	N/A
Chilliness	29.3	0.43
Any hot flush	**22.1**	
*Hoteri**	17.1	0.6
Low libido	12.2	0.51
*Kyū na nekkan**	10.7	0.64
*Nobose**	9.3	0.69
Sudden sweating	8.6	0.48
Anhedonia	8.6	0.47
Night sweats	6.4	N/A
Vaginal dryness	5	0.77
Hotto furasshu	3	N/A
Painful/difficult sexual intercourse	2.9	0.76

* Shaded rows indicate Japanese terms Lock used to gloss hot flush in her research.

I collected data on the frequency of symptoms and then conducted factor analysis. I was interested not only in assessing whether symptoms of *kōnenki* were different from those of Western biomedical menopause, but also whether Japanese women (particularly those consuming higher quantities of soy phytoestrogens) had lower rates of these symptoms. Factor analysis for the full list of symptoms is described in detail in Chapter 6 on the proximate culture of aging. Here I draw from the same study to focus more specifically on symptoms which are likely to share an underlying hormonal etiology including vasomotor symptoms such as hot flushes.

My plan was to conduct factor analysis using data from a checklist of symptoms recalled in the previous two weeks. I predicted that the three hot flush terms used by Lock in her research (i.e., *kyū na nekkan, nobose, hoteri*) would all be grouped in the same factor, thereby justifying lumping them together by Lock and by me in the daily symptom checklists. The results of factor loadings on what I called a sexual-vasomotor (SV) symptom factor are shown in Table 7.1, in descending order by frequency. The second column includes the factor loadings when analyses were conducted using symptom frequency (see reference for analyses using frequency-severity scores).[5]

The foreign loan word *hotto furasshu* was not in that factor, but rather in a psychological-anxiety factor when frequency data were used, and in a psychosomatic factor when frequency-severity data were used. This factor analysis also noted other surprising results. Night sweats, understood as a common vasomotor symptom in many other studies worldwide, was not in the sexual-vasomotor factor, whether using frequency or frequency-severity data. In contrast, chilliness made a surprise appearance in this sexual-vasomotor factor, suggesting a robust, albeit unexpected, association with hot flushes. Of all the symptoms in the sexual-vasomotor factor, chilliness also had the highest frequency of occurrence, suggesting that it was an important symptom, reported by almost a third of the study participants.

7.2.3 Hot Flushes

Cross-cultural studies often challenge Western outsider assumptions about what constitutes "normal." Lock's research on menopause in Japan and North America led her to propose that there were many cultural factors influencing the differences between the local biologies and experiences of menopause in the United States, Canada, and Japan.[6] By examining Japanese women's experiences of aging and menopause, Lock wanted to decenter the "normal" biomedical white body, the standardized body for which disease states are defined and medications are developed and prescribed. Lock said that she almost never heard a Japanese woman complaining of a hot flush. Whereas when she did interviews with women in North America, she sometimes saw them having hot flushes, dripping with sweat. She concluded that there was clearly something different going on in these two populations. Despite Japanese recognizing more variation in hot flushes (i.e., more linguistic marking, suggesting more awareness of nuances in physiology and experience), Japanese women had lower rates of hot flushes compared to their North American counterparts, no matter which terms were counted (or even when all terms were combined).[6]

In our study, in the two-week recall of symptoms (which tends to be the standard recall period because people have trouble remembering longer), when I separated the hot flush terms, response rates gave an over seven-fold range of prevalence depending on the term used – in the same population sample. Imagine the difficulties of comparing studies that use different terms! How do we even begin to assess the true rate of menopausal hot flushes in any population? If I had simply asked about *hotto furasshu*, I would have thought only 3% of women experienced hot flushes. This compares to a rate of 22.1% if any hot flush is included (all three shaded rows in Table 7.1). How can what appear (to English speakers) simply to be linguistic differences yield such varied results or be so different? Are they different phenomena or different flavors of hot flushes?

On the daily symptom checklists, I followed Lock's lead and grouped the three hot flush terms together (*kyū na nekkan*, *nobose, hoteri*) in a single (albeit larger) row to gloss the concept. However, after receiving the first few weeks of daily symptom checklists in the mail, I discovered that several women had responded separately for *kyū na nekkan, nobose,* and *hoteri.* While a few participants did this for the other symptoms I lumped together in a single row (e.g., joint pain and muscle pain), most were okay with answering those with a single entry in the row that listed two terms. This was not the case with the hot flush terms.

These women clearly thought I had made a mistake on my form and that there should have been three distinct rows where there was only one. From an insider perspective, these were different concepts – that's why there were different words! Duh! My mistake reflected my Western biomedical (outsider) bias that hot flushes were a single symptom and thus this bias did not allow for nuance or distinction. I wonder if I had drawn lines between the symptoms to create three distinct rows, how many other participants might have reported differently on the daily symptom checklists! This was a lost opportunity to analyze how other factors (e.g., diet) might influence different types of hot flushes. Furthermore, it demonstrates how research is both art and science, and that participants have a lot to teach us about the biases inherent in our "objective" outsider/*etic* lens!

Fortunately, I had separated the three hot flush terms for the two-week recall checklist, which enabled me to examine them separately in factor analysis. But because I wanted to avoid biasing participants by disclosing that I was really interested in menopause, I waited until the final interview to ask more probing questions about how the various hot flush terms differed. And when I did ask, I got more "thick description" (context)[7] than I could ever have imagined.

The seven-fold range of prevalence depending on the hot flush term got me interested in other words that might be used for hot flushes, and how they might differ in terms of meaning, whether biologically or culturally. Fieldwork highlighted that the terms used in the initial symptom checklist may have been too limited. During the final interviews, I probed women further on hot flushes, asking them about all the terms I had heard in conversation or read in the literature.

There were six terms in total: the three that Lock used to gloss hot flush in her research; fourth the relatively new foreign loan word *hotto furasshu* I had included in my two-week symptom checklist; fifth, a somewhat generic *atsuku naru* which simply means to become hot; and sixth, the onomatopoeic expression, *kaa to atsuku naru*, that conveys a sudden feeling of heat.[8] Because *hotto furasshu* was reported by so few people, sorted into a different symptom factor, and seemed not quite related to the idea of a menopausal hot flush, it was excluded from subsequent analyses. The relative frequency of use of terms among respondents is presented in Table 7.2.

TABLE 7.2 Cultural salience of hot flush terminology: Percentage of participants ($N = 117$) able to define each hot flush term

Symptom term	*Response frequency (%)*
Kaa	87.1
Hoteri	77.0
Nobose	76.3
Atsuku naru	64.0
Kyū na nekkan	55.4

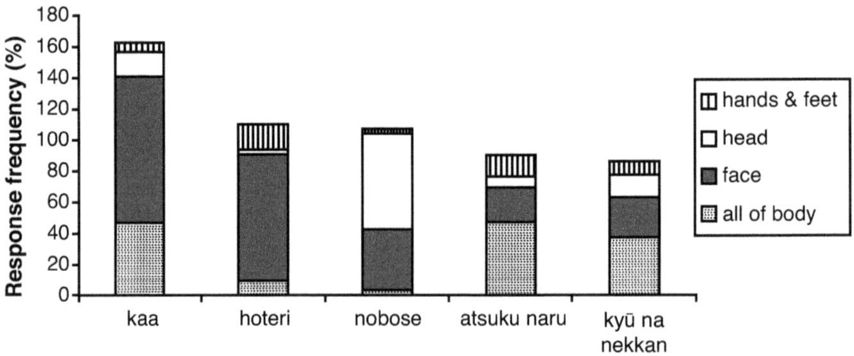

FIGURE 7.2 Specificity of hot flush terminology: physical location of symptom reported by participants (% of positive responses). Response frequency is >100% in some cases because some people reported more than one location per term (reproduced with permission).[2]

Participants explained how different terms conveyed feelings of heat on different parts of the body (mapped in Figure 7.2). Many said that *hoteri* is experienced mostly on the face, secondarily on the hands and feet. *Nobose* is mostly experienced on the head, and secondarily on the face. Some of the other words like *atsuku naru* and *kyū na nekkan* were less specific, with most people saying they referred to feelings of heat over the entire body. The onomatopoeic phrase *kaa to atsuku naru* primarily referred to heat on the face.

I asked people to describe not just the part of the body, but the sensation and other characteristics of each term. For *nobose*, people often said it entailed feeling dizzy and woozy, like the kind of heat you get from sitting too long in a hot bath, drinking alcohol, or being in crowds. *Hoteri* seems to be more of a superficial feeling, experienced more on the skin surface. *Kaa to atsuku naru* tends to feel like heat rising, sometimes followed by cold and/or sweating. Might some people feel more bothered by the feeling of cold/chill after a hot flush, and notice different attributes of a hot flush? Interestingly, *kaa to atsuku naru* also had a very important psychological component of feeling nervous, irritable, and angry. *Atsuku naru* and *kyū na nekkan* were the most generic, reportedly affecting the entire body according to most respondents. Compared to English, Japanese appears to have a much more descriptive language for hot flushes.

Does this terminology difference mean there are real differences in body locations where hot flushes occur, or do different people notice hot flushes on certain areas of their bodies more than other parts of their bodies? Are these differences simply spatial or could there also be a temporal component? Kronenberg took thermal photos of hot flushes as they occurred in North American women and documented the temporality of hot flushes. In a hot flush lasting several minutes, different regions of the body appear hotter at different time points (Kronenberg, pers. comm.). Since *nobose* occurs on the head and face, and *hoteri* occurs on the face and hands, could these terms be used to refer to different time points in a hot flush? Is it possible that *nobose* refers to the initial stage and *hoteri* to the middle stage of a hot flush? Do people pay attention to different stages when they label their hot flushes? Or is there variable physiology? As with much ethnography, digging deeper into phenomena (such as hot flushes) often raises more questions than it answers.

7.3 Other Vasomotor Symptoms

One of the more interesting findings from the data collected on VMS was not only that there are many different terms for hot flushes and variation in what they refer to, but also the existence of the symptom of chilliness (*hieshō*) and its association with established VMS like hot flushes. I was surprised to see chilliness in the sexual-vasomotor factor, albeit with lower factor loadings than sexual and hot flush symptoms – and began to question whether it should be considered a vasomotor symptom. My observation on chilliness served as a reminder of the importance of being open-minded and listening to people to discover potential "new symptoms" that don't seem to exist in researchers' culture, or at least have no explicit label.

7.3.1 Criteria for Vasomotor Symptoms (VMS)

To be considered a vasomotor symptom of menopause, a symptom should have: a statistical association with other symptoms assumed to have an underlying hormonal basis (e.g., shown by factor analysis described in Chapter 6), a temporal profile associated with a peak during the menopausal transition, and a potential mechanism related to changes that occur with menopause.

Importantly, chilliness was the most prevalent (frequently reported) vasomotor symptom in the sexual-vasomotor factor. This was not a typical or expected vasomotor symptom. Chilliness is not part of the constellation of typical menopausal symptoms for the "normal" white Western human body.

Thus, it was thus not expected by me coming from the West with my biases. Yet, 29.3% of the study population reported experiencing it in the previous two weeks. This compared with a range of 3.0–22.1% for hot flushes, suggesting that if chilliness is a vasomotor symptom, it is the most prevalent vasomotor symptom in this population.

Chilliness exhibited statistically significant ($P < 0.001$) positive correlations with other typical vasomotor symptoms such as hot flushes (*nobose* and *hoteri*) and sudden sweating. This means that women who had hot flushes and sweats were also more likely to experience chilliness. Chilliness prevalence differed significantly ($P < 0.01$) between premenopausal (reported by 11%) and peri- and postmenopausal status groups (reported by 35–38%). This suggests some disruption in the thermoregulatory system, possibly due to shared underlying changes of the menopausal transition.

Interestingly, though, significant differences in chilliness were not found by *kōnenki* category. In contrast, significant differences in *hoteri* and sudden sweats were observed by *kōnenki* category. This suggests that while *hoteri* and sudden sweats may have contributed to *kōnenki* stage classification, chilliness was not a key symptom used for self-identification of *kōnenki* status. Particularly noteworthy for a proximate cultural perspective on *kōnenki* and menopause, the insider/*emic* symptom of chilliness appears to correlate with the outsider/*etic* categories of menopause, but not the insider/*emic* categories of *kōnenki*!

Could chilliness be a more important vasomotor symptom than hot flushes and sweats in Japanese women? Do we need to expand our views of thermoregulation to include both ends of the spectrum to include both hot flushes/sweating and chilliness? Mechanistically, hot flushes involve thermoregulation, and for Western biomedicine, the focus has been on the sweating threshold. Chilliness also involves thermoregulation, but at the opposite end of the spectrum, one which has not been studied in the context of Western biomedical menopause.

7.3.2 Potential Mechanisms for Chilliness

Cultural factors such as diet could influence potential mechanisms, which may operate through body composition and dietary components. One potential explanation for chilliness among Japanese women is that Japanese women's relative thinness might make them more likely to feel cold. Researchers in the United States have found that BMI often correlates with hot flushes, and the "insulation hypothesis" suggests that higher BMI correlates with higher "insulating" fat which may trigger or exacerbate hot flushes. However, fat also provides sources of peripheral estrogen (estrone) which may act like HRT to decrease the likelihood of hot flushes ("peripheral estrogen hypothesis").

Data from this study in Japanese women found that increased BMI (for the most part still in "healthy" ranges below cutoffs for overweight or obesity), correlates positively with hot flushes (*nobose*), sudden sweating, and night sweats. This supports the "insulation hypothesis" for hot flushes and leads to the hypothesis that if chilliness is at the opposite end of the thermoregulatory spectrum, BMI should be negatively correlated with chilliness. However, no such significant correlation was observed, and thus, the insulation hypothesis does not seem supported in the case of chilliness, at least in this sample.[9] However, dietary soy phytoestrogens could offset the lack of estrogen from body fat.

Could culture in the form of dietary soy phytoestrogens play a role in the patterns of VMS (including chilliness) observed in Japan, thereby providing a (bio)cultural mechanism influencing the local biology of Japanese women? Evidence exists for soy phytoestrogen intake being negatively correlated with hot flush frequency and severity.[10] By extension, if chilliness is a "real" vasomotor symptom, we might hypothesize that it would be negatively correlated with phytoestrogen intake or exposure. Negative correlations between soy phytoestrogen intake measured by dietary records were observed for sudden sweating and night sweats, and also between chilliness and dried blood spot soy phytoestrogen exposure.[9] This means that women with higher intake of and exposure to soy phytoestrogen were less likely to report sweats and chilliness respectively. Interestingly, no correlations were found for soy phytoestrogens and the hot flushes *nobose* and *hoteri*, although they were more prevalent than sudden sweating and night sweats (which did show significant inverse relationships with soy phytoestrogen intake). Clearly, more needs to be investigated here.

A negative correlation between soy intake and/or exposure and VMS is usually interpreted to mean that eating more soy will decrease VMS. However, correlations do not allow us to conclude anything about causation. There is evidence from experimental studies that intake of soy phytoestrogens leads to decreased frequency and severity of VMS like hot flushes.[10] However, I have never heard anyone talk about eating soy to decrease chilliness. Could chilliness among Japanese women reflect differing thermoregulatory physiology and setpoints, possibly influenced by lifelong dietary soy phytoestrogen exposure?

It is one thing to report correlations between symptoms and diet, but another to hypothesize potential mechanisms. With hot flushes, there is evidence that phytoestrogens may function similarly to estrogen replacement therapy to decrease frequency of hot flushes (although the mechanism is not really understood for either HRT or phytoestrogens). But how might phytoestrogens work for chilliness? Since no one had published on chilliness as a vasomotor symptom before our article appeared in 2007, no one had investigated whether HRT could treat it. Even though Freedman had extensively studied factors influencing the sweating threshold of the thermoneutral zone, and the occurrence of factors influencing hot flushes, when I asked him about the effects on the shivering threshold, he told me he had never investigated it because nobody had ever complained of chilliness in his Western sample population. Our observations and hypotheses[9] need to be investigated further, and I hope that readers of this book will be inspired to do so.

7.3.3 Local Biology of Chilliness in Japan

In Western biomedicine, the focus has been on hot flushes, which appear to be on one end of a continuum of thermoregulation. Chilliness appears to be on the opposite end. Could an expanded awareness of relevant thermoregulation symptoms, including chilliness, increase our understanding of changing physiology during the menopausal transition? Could it also help us look beyond potentially simplistic explanations involving a single symptom and single hormone? Researchers and clinicians in the United States have focused almost exclusively on estrogen, which may have hampered understanding of the complexity of changes during menopause.

Perhaps the traditional Japanese focus on the autonomic nervous system as playing a role during *kōnenki* might provide clues to the etiology of hot flushes, as well as chilliness. The autonomic nervous system regulates sympathetic activity. The stable body temperature zone between the sweating and shivering thresholds (thermoneutral zone) is reduced in women with symptoms of hot flushes.[11] Could the same be true for women with chilliness? Relaxation exercises, which influence sympathetic activity, have resulted in decreased hot flushes,[11] possibly by expanding the thermoneutral zone.[12] Could interventions that decrease hot flushes by increasing the thermoneutral zone also work to decrease chilliness? Since soy phytoestrogen exposure is inversely correlated with chilliness frequency, this suggests that phytoestrogens may act to widen the thermoneutral zone, giving the body a wider range of core body temperature stability.

When I submitted a paper arguing that chilliness appeared to be an important vasomotor symptom among Japanese women, one journal reviewer initially rejected it, claiming that chilliness was not a vasomotor symptom, presumably basing this on their clinical experience and knowledge of the biomedical literature. However, my research in Japan documented that chilliness: is more prevalent than hot flushes, correlates with other well-established VMS like hot flushes, has factor loadings on a vasomotor factor suggesting shared underlying etiology (e.g., hormones), showed significant negative correlations with phytoestrogen exposure, and has a purported vasomotor mechanistic basis. Thus, at least in Japan, it should be considered a vasomotor symptom. Beliefs about

VMS may not be completely "scientific and objective," but rather influenced by the culture, sensibilities, perceptions, and environments in which they have been developed (even if they are assumed to be "neutral" or "standard"). The journal reviewer did not look at the evidence I presented but rather seemed to respond from their own Western biomedical biases, having not heard of chilliness as a vasomotor symptom or even a symptom at all. By proposing to add a new symptom to the category of VMS, my work challenged the reviewer's notion of what was contained in that category, and the response seemed to highlight the clash of the Western biomedical outsider/*etic* and Japanese insider/*emic* systems. In the end, I convinced the editor that my data deserved to be published so that others could expand their view of potential symptoms and begin to examine this issue further. But I wonder how many similar stories exist, where authors with data that challenged dominant biomedical worldviews were unable to publish their results, leading to increasingly homogeneous reinforcing narratives about how health and human bodies work.

Part II demonstrates how listening to what people say (not just words, but definitions and nuances) can shed light on proximate cultural aspects of aging and menopause and can lead us down novel paths to more expansive views of lived experience and local biology. There are several take home lessons from examining the experience of and language used to convey the experience of *kōnenki* in Japan.

1 Interpretation of epidemiological prevalence estimates requires understanding of cultural/linguistic variation, since we observed a seven-fold variation in hot flush rates depending on the term used. Language is critical when comparing health-related issues cross-culturally.
2 It is critical to avoid *a priori* assumptions about what constitutes menopausal or even vasomotor symptoms, since the latter was found to include chilliness among Japanese midlife women. This symptom is not generally recognized in the West, but once I published the paper on chilliness and began presenting on it, women from Western countries started approaching me after talks and telling me that they or someone they knew had experienced such symptoms. So perhaps our biases have prevented us from gaining a more complete understanding in the West. Interestingly, the insider/*emic* symptom of chilliness appears to correlate with the outsider/*etic* categories of menopause, but not the insider/*emic* categories of *kōnenki*! This reinforces the earlier observation that *kōnenki* and menopause do differ significantly.
3 We need to consider comparison groups or controls, so we are not limited by the research window, because men also experience many *kōnenki* symptoms and these data are critical for us to distinguish between symptoms of aging vs. "menopause."

In Part II we broadened our perspective on aging and menopause by expanding cross-culturally across space, from Western biomedical culture to Japanese culture. In Part III, we broaden our perspective in terms of time. We take a long-term evolutionary approach to ask the question, "why do aging and menopause exist," and to consider whether aspects of our modern (and Western) lifestyles may play a role in how aging and menopause are experienced today.

References

1. Kaufert PA. 1982. Anthropology and the menopause: The development of a theoretical framework. *Maturitas.* 4(3):181–193. https://doi.org/10.1016/0378-5122(82)90048-2
2. Melby MK. 2005. Vasomotor symptom prevalence and language of menopause in Japan. *Menopause.* 12(3):250–257. https://journals.lww.com/menopausejournal/abstract/2005/12030/vasomotor_symptom_prevalence_and_language_of.5.aspx
3. Melby MK, Lock M, Kaufert P. 2005. Culture and symptom reporting at menopause. *Hum Reprod Update.* 11(5):495–512. https://doi.org/10.1093/humupd/dmi018
4. Park J, Johantgen ME. 2017. A cross-cultural comparison of symptom reporting and symptom Clusters in heart failure. *J Transcult Nurs.* 28(4):372–380. https://doi.org/10.1177/1043659616651673
5. Melby MK. 2005. Factor analysis of climacteric symptoms in Japan. *Maturitas.* 52(3–4):205–222. https://doi.org/10.1016/j.maturitas.2005.02.002
6. Lock M. 1993. *Encounters with Aging: Mythologies of menopause in Japan and North America.* Berkeley, CA: University of California Press.
7. Luhrmann TM. 2001. Thick description: Methodology. In: Smelser NJ, Baltes PB, eds. *International Encyclopedia of the Social & Behavioral Sciences.* Oxford: Pergamon; 15665–15668. https://doi.org/10.1016/B0-08-043076-7/00769-5.
8. Zeserson JM. 2001. How Japanese women talk about hot flushes: Implications for menopause research. *Med Anthropol Q.* 15(2):189–205. https://doi.org/10.1525/maq.2001.15.2.189
9. Melby MK. 2007. Chilliness: A vasomotor symptom in Japan. *Menopause.* 14(4):1–8. https://doi.org/10.1097/gme.0b013e31804ffd81
10. Taku K, Melby MK et al. 2012. Extracted or synthesized soybean isoflavones reduce menopausal hot flash frequency and severity: Systematic review and meta-analysis of randomized controlled trials. *Menopause.* 19(7):776–790. https://doi.org/10.1097/gme.0b013e3182410159
11. Freedman RR. 2014. Menopausal hot flashes: Mechanisms, endocrinology, treatment. *J Steroid Biochem Mol Biol.* 142:115–120. https://doi.org/10.1016/j.jsbmb.2013.08.010
12. Freedman RR. 2005. Hot flashes: Behavioral treatments, mechanisms, and relation to sleep. *Am J Med.* 118(12 Suppl 2):124–130. https://doi.org/10.1016/j.amjmed.2005.09.046

PART III
Ultimate Biology

8

ULTIMATE BIOLOGY

Introduction

Part III on the ultimate biological quadrant of the framework introduced in Chapter 1, focuses on *why* biological phenomena happen (Figure 8.1). Specifically, why do we age? Why can't humans live forever? Why does menopause exist? Why do women live longer than they can reproduce? Is menopause somehow adaptive or functional? These questions involve longevity and life expectancy, and we need to understand these terms before we can examine the traits that pose evolutionary puzzles. This chapter reviews evolutionary theory to explore why people age, and why menopause happens.

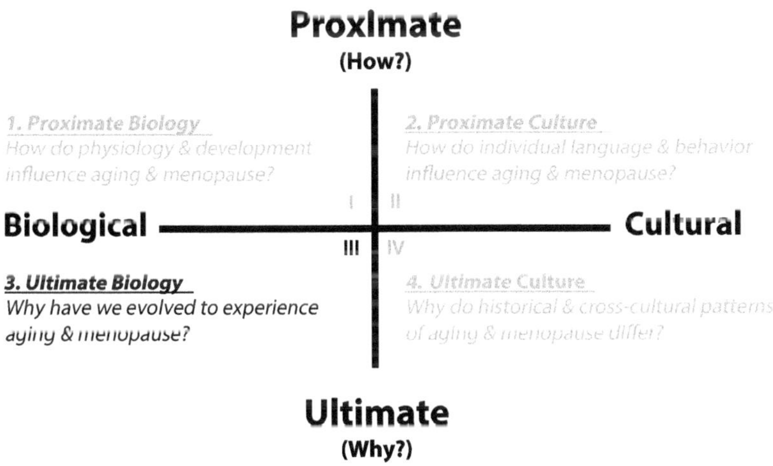

FIGURE 8.1 Ultimate Biology Quadrant III of the Proximate-Ultimate Biocultural Framework introduced in Chapter 1.

DOI: 10.4324/9781003644521-11

8.1 Life Expectancy

The New Year is arguably one of the most important holidays in Japan. Everyone sends *nengajo* (New Year's postcards), which are delivered on New Year's Day. Until email became ubiquitous, this was probably the most lucrative holiday for the Japanese postal system. At its peak in the mid-2000s, billions of cards were sent to family, friends, colleagues, and business associates – anyone people felt indebted to and wanted to thank or update on their lives.

Fieldnotes

My first New Year in Japan, I sent nengajo *(New Year's postcards) to my study participants and collaborators. I wrote them by hand, painstakingly looking up many* kanji *(Japanese characters) and copying them. Since none of my participants expected* nengajo *from me, they had not sent any to me. But if you receive a* nengajo *from someone whom you did not send one to, you are supposed to send one during the* matsunouchi *period (one to two weeks following New Year's) and update your list for the following year. Over the years, my list grew and so did the number of cards I received.*

In a year when a close relative dies, a person does not send out nengajo, *but instead sends out* mochū hagaki *(mourning postcard) to inform others that they are in mourning and to share information, including age, about the deceased. Within a couple of years of moving to Japan, as my list grew to over 150 (including participants, colleagues, and friends), I began to receive such cards – which I had never seen before. I asked friends, how should I respond in such a case? Their answer: don't mail* nengajo *to that person that year, but instead after the* matsunouchi *period, send* kanchū mimai *(mid-winter card), explaining that you were aware they were in mourning (which is why you had not sent a regular* nengajo*) and offering condolences.*

From a population-level public health perspective, the fact that most of the *mochū hagaki* I received concerned people in their 90s seemed like something to celebrate! My collection of these cards aligned with the claim that Japanese (especially women) have the longest life expectancy in the world.[1] And since life expectancy is a measure of population health, these cards would seem to support the idea that Japan is doing something right!

These Japanese women who died in their 90s spent more time being post-reproductive (40+ years, assuming an average age of 50 for menopause) than they did being potentially reproductive (25–35 years depending on when they became biologically and socially able to become pregnant). This seems odd from an evolutionary perspective. Why would any animal spend so long being post-reproductive? If this is a modern phenomenon due to recent increases in life expectancy caused by contemporary cultural developments, then speculating about evolutionary benefits is unnecessary, because it rarely occurred in the past. But is this the case? Were there really few post-menopausal women in the past? Demographic reconstructions suggesting ancestral human life expectancies of less than 50 years have led to a view that menopause is a relatively recent phenomenon in human history. But we need to evaluate this claim, and to do that, we need to establish a few definitions.

Long life (longevity) is obviously relative and exhibits individual variation, with greater longevity on average observed in women compared to men and in populations with higher socioeconomic status or healthier lifestyles (see Part IV). Life expectancy is the population average for the remaining years of life expected at a given age. Life expectancy at birth is a widely used measure for assessing population health because it incorporates mortality throughout the life course by giving us the average age of death in any population. Before modern hygiene, nutrition, and medicine, life expectancy was estimated to be about 30 years, and then more than doubled to around 70 years by 2021.[2] While life expectancy is a useful quantitative measure of population health, health also has important qualitative components that are harder to measure.

8.1.1 Misunderstandings about Averages

When evaluating life expectancy, there is a common misunderstanding about averages that is relevant for our discussion of the evolution of aging and menopause. While "life expectancy" usually refers to "life expectancy at birth," it is possible to calculate life expectancy from different ages, in effect estimating how long the average person of a certain age can expect to live.

Since infant mortality has been high in many human populations throughout time, and still is in many parts of the world, the average life expectancy at birth is lowered considerably by those infants and young children who die in the first few years of life. If a woman survives until adulthood, her likelihood of surviving past menopause is increased, compared to her life expectancy at birth. Thus, while life expectancy at birth may have been low for much of human history, women who reached adulthood or survived their reproductive years could have expected to live well past menopause.

"Average life expectancy" can be very misleading, making it difficult to understand the range of individual experiences in a population. Consider the following examples of small four-person "populations," all with average life expectancy of 40 years, yet with underlying data that reflect vastly different experiences for individual population members.

- *Example A:* Ages at death: 6 months, 18 months, 78 years, and 80 years. Sum of ages at death/# of dead = 160/4 = 40 years
- *Example B:* Ages at death: 38, 39, 41, and 42 years. Sum of ages at death/# of dead = 160/4 = 40 years
- *Example C:* Ages at death: 10, 30, 50, and 70 years. Sum of ages at death/# of dead = 160/4 = 40 years

Thus, to know that each of these populations has a life expectancy at birth of 40 years tells us very little about the experience of death of individuals, or the likelihood of living past age 50, in each population. Generally, populations with low life expectancies at birth experience high infant and child mortality. Every child who dies at three days of age, or one week, or one month, pulls the average age of death down dramatically.

Taking a life course perspective on mortality, in most populations an individual's chance of dying starts relatively high at birth, but it drops considerably after infancy and reaches the lowest point right around puberty. After puberty, when reproduction becomes possible, the chance of death starts increasing, due partly to accidents and injuries, but also to increased risk of disease,[3] likely resulting in part from tradeoffs of investing in reproduction instead of maintenance and repair of the body. Since the 1950s, a new mortality trend has begun in the form of unexpected declines in mortality rates at advanced ages and "rectangularization" (or flattening of the curve with more sudden drop off near the end) of the survival curve. This development is due in large part to medical interventions which extend life expectancy, but sometimes at the expense of quality of life.

8.1.2 Birth Rates and Implications for Detecting Menopause

High infant mortality rates among ancestral populations (and contemporary hunter-gatherers and populations living in poverty) often go hand in hand with higher fertility, and these have implications for aging and menopause. Symptoms of aging and menopause may be more likely to be noticed in current populations, not only due to recent increases in life expectancy, but also because of lifestyles and reproductive patterns which have departed considerably from those of our ancestors.

Fewer pregnancies and shorter duration of lactation in industrialized populations result in greater numbers of menstrual cycles over the life course, and thus perhaps heightened awareness of menstruation and its end. Compared to contemporary industrialized populations, women living in "natural fertility populations" without birth control may experience many pregnancies and breastfeed for years following each birth, leading to relatively few periods over the life course. In traditional (natural fertility) populations, a woman may have her last pregnancy in her mid-40s, followed by breastfeeding for several years. Thus, she may enter perimenopause at a time when she is not having periods and may not even notice she has gone through menopause.

8.1.3 Can Lifespan Be Increased?

The length of time a person or species can live (lifespan) can range from over 100 years in terrestrial animals such as tortoises to thousands of years for some trees and some aquatic animals like sponges. Humans have the longest lifespan among terrestrial mammals, but several marine species such as the bowhead whale have lifespans considerably longer than ours. For comparison, common animal models such as lab mice have lifespans of two to three years, and fruit flies have lifespans of only about 50 days.

Lifespan is influenced by environmental factors that influence the risk of mortality and can be extended with interventions like caloric restriction (see Part I), although often not without costs involving immune function and stress responses.[4] Ultimate biological questions exist about whether the human lifespan has a limit (currently estimated at 125 years), and answering such questions requires concepts of evolution (e.g., selection, adaptation) and Life History Theory.

Darwin's Theory of Evolution by Natural Selection, a process by which organisms that are better adapted to their environments survive and reproduce more, is one of the most paradigm-shifting theories in biology. Darwin (and contemporaries) did not know about DNA or genes, but he did understand heredity, and people had been using that understanding for artificial selection and domestication of plants and animals for thousands of years. Thus, thinking on an ultimate biological or evolutionary level does not require knowledge of genetics or molecular biology. Understanding some basics of evolution allows us to delve deeper into evolutionary hypotheses about aging and menopause.

8.2 Evolutionary Theory

Evolution is the source of most of the incredible biological diversity on the planet, with mutation and sexual reproduction contributing genetic diversity for selection to act up on. Evolution is assessed by examining traits (or gene frequencies) on a population level, but the processes occur by differential survival and reproduction of individuals. Individuals' unique DNA (genotypes) interact with the environment over the life course to produce observable physical properties, including appearance and behavior (phenotypes). The changes involved in evolution can occur in four ways, through: genetic mutation, genetic drift, gene flow, and selection (which is often divided into natural, artificial, and sexual selection).

8.2.1 Evolutionary Basics

First, a quick biology review (or more[5]). Genes are composed of DNA, ranging from a few hundred to a few million base pairs long. DNA gets transcribed by messenger RNA and then translated into protein molecules. Genes can have different variants (alleles) that may vary by a single base pair or by many. Some of the variation matters (and may change the molecule coded for by the gene); some does not, because of redundancy in the DNA-to-protein code. For example, several different combinations of three DNA base pairs (codons) can be translated into messenger RNA molecules that code for the same amino acid, the building blocks of proteins.

Evolution is defined as the change in allele frequencies in a population over generations. Several processes affect allele frequencies in populations. Genetic variation is required for selective forces to act upon and is increased by mutations (changes in DNA sequence due to random errors in DNA replication, exposure to mutagens, or viral infections). Genetic mutation is like the childhood game of telephone where children stand in a circle, one person starts a message and transmits it to the person next to them and so on, and over time, the message gets changed, often significantly. Mutation is also akin to the mistakes made by monks when copying texts before the invention of the printing press. Mutation creates the variation that selection acts upon. Most mutations are neutral, some can be detrimental to survival and reproduction, and a few can be advantageous for survival and reproduction (adaptive) in specific environments.

Gene flow (e.g., migration) can increase variation in a population. Genetic variation can also increase by crossing-over at meiosis, recombination, and sexual reproduction. In sexual reproduction, offspring receive one allele from each parent. Thus, for a complete understanding of the evolution of most traits in species that sexually reproduce, examination of both males and females may be necessary. This is relevant to understanding the evolution of menopause.

Bacteria are particularly good at exchanging DNA (without having sex) through a mechanism called horizontal gene transfer. And bacteria can reproduce quickly (i.e., have short generation times) compared to humans. While humans may have more technological abilities in terms of medical treatments and vaccines, short bacterial generation times give pathogenic bacteria a potential competitive advantage over humans. But mutualistic bacteria may also give humans adaptive benefits, allowing them to adapt quickly to changes in the environment, including their diets and other microbes.

Genes can be selectively turned on or off by molecular "switches" (epigenetics), influenced by environmental factors such as diet, and these switches appear to be partly heritable, although still not well understood.[6] Genetic variation can be decreased by bottleneck effects and migration causing founder effects, as occurred in Amish and Ashkenazi communities (genetic drift). Selection also decreases genetic variation in populations.

8.2.2 Beyond (Host) Genes

Adaptation occurs when individuals with particular traits (phenotypes) have higher reproductive fitness than other individuals in certain environments. Phenotypes are produced by the interaction between genes (genotypes) and the environment, with environmental factors influencing which genes are expressed and when, as represented in the equation:

Phenotype = Genotype + Environment

But host organism genes are only part of the equation. Human bodies are composed of roughly equal numbers of human cells and single-celled microbes, and many more microbial genes (microbiome) than human genes.[7] The human microbiome can be considered as both part of us and also part of the environment. Thus, we may need to rethink the standard equation of Phenotype = Genotype + Environment, to include genotypes of the myriad microbes residing within and on us, and/or recognize that microbes (e.g., of the gastrointestinal system and the skin) mediate critical interactions with the environment. For example, diet is a major environmental factor influencing health, but diet is mediated by the gut microbiome, without which we would not have many essential vitamins and chemicals critical for our survival, development, and health. As microbes have a much shorter generation time compared to humans, the microbiome may allow faster adaptation to environmental changes (e.g., diet). Thus, our microbiome may be part of the ecological adaptation and success of humans on the planet.

Epigenetics is another phenomenon influencing how genes and environment interact to produce variation in phenotypes. Environmental factors can turn genes on and off after exposures, and the chemical markers that do this can be passed down across multiple generations. Similarly, microbes (and their genes) can also be passed from mother to offspring through birth and perhaps in mammals through nursing.[8] Previously, we thought only genes could be inherited. But the lines between Genotype and Environment in the above equation become blurred when we recognize that environmental epigenetic markers can be heritable, and that some of our key genes are actually microbial and also may be heritable. Thus, the microbiome might extend host evolutionary potential[9] and provide other genes upon which selection may act.

8.3 Selection

Selection is often divided into three types: natural, artificial, and sexual. In artificial selection, humans are the selecting agent. And although humans are part of nature, since we create the categories, we have put ourselves into a different category when acting as selective agents. In sexual selection, potential mates are the selecting agents, and may select for traits that are not adaptive for survival, thus possibly creating tension between natural and sexual selection, although some theorists have subsumed sexual selection under natural selection. Artificial selection could be subsumed under natural selection, if humans were to consider themselves as part of nature.

8.3.1 Natural Selection

Ultimate biological explanations about aging and menopause examine whether key traits are beneficial to survival and reproduction (adaptive) in certain environments and have been subject to selection. The "key ingredients" of natural selection are: phenotypic variation, heredity, competition, and differential survival and reproduction, and heredity. In other words, there must be variation in phenotypes that leads to differential survival and reproduction, and the variation in phenotypic traits must be heritable. Although scientists often anthropomorphize selection, making it sound more intentional and directional, it is in fact mindless and mechanistic. People often think that organisms evolve certain traits in response to selection, but it is important to keep the order straight. First, a random mutation arises in an individual, generating variation that may lead to differential reproductive success (or fitness, the currency in evolution). If a mutation confers advantage (in terms of reproductive success), then more of the next generation will have that mutant allele, and over time, more of the population will have that mutation (i.e., the allele frequency will increase in the population). We then say natural selection has "acted" to select for that variant allele, although intentionality and direction do not exist – it is dependent on the environmental context. There is not an "invisible hand" of natural selection reaching down and picking off organisms without a particular gene and choosing ones with a particular gene, but rather gene frequencies simply shift over time due to differential survival and reproduction of organisms with different traits in particular environments. Variation must *already be there* for selection to act upon. Selection *reduces* variation; it does not *create* variation. Successful variations accumulate and eventually lead to different species.

Charles Darwin and Alfred Wallace independently thought of the Theory of Natural Selection without knowing about genes; rather they focused on traits. We often don't know which genes contribute to specific traits, and even identifying the trait can be challenging – particularly with phenomena like aging and menopause.

Natural selection generally occurs on the level of the individual, not on the level of groups although debates on group selection continue. While contemporary evolutionary discussions tend to focus heavily on genes, we don't need to have identified genes to use the Theory of Natural Selection and think about phenomena evolutionarily. Darwin didn't.

Darwin's Theory of Natural Selection has four components:

1 *Variation*: Individuals within species have variation in traits (e.g., height, disease resistance)
2 *Inheritance/Heritability*: Traits (or genes coding for them) are passed from parent to offspring
3 *Competition*: Most populations have more offspring than can be supported by the environment, creating competition (e.g., for food but also mates) and a "struggle for existence" resulting in:
4 *Differential survival and reproduction*: Those offspring with traits more adaptive to their current environment are more likely to survive and reproduce

Natural selection occurs when non-human selective agents (e.g., climate, other animals) lead to differential survival and reproduction. Artificial selection occurs when humans act as selective agents choosing plants and animals with desired characteristics (e.g., domestication). For example, humans have selected certain varieties of plants and certain breeds of domesticated animals (for food as in cows and chickens, and companionship as in cats and dogs). Sometimes, humans also unintentionally select certain varieties of organisms.

8.3.2 Artificial Selection

Let's consider an example of unintentional artificial selection that you may be familiar with, as it is probably one of the biggest public health challenges facing humans today: antibiotic resistance.[10] The use of antibiotics, for treating or preventing infections, also selects for antibiotic-resistant bacteria in whatever environment they are used (e.g., human bodies and animals we eat), posing risks not only to individuals using antibiotics but to the larger population.

Human bodies are home to many microbes that comprise the human microbiome, some of which have mutations that confer resistance to specific antibiotics. Bacteria with mutations that allow them to escape the effects of an antibiotic (e.g., by preventing it from entering their cells, pumping it out, or disabling it), are called "resistant bacteria," and will be able to exploit new ecological habitats (niches when antibiotics kill other susceptible bacteria. Thus, an antibiotic may select for resistant bacteria, allowing them to increase their frequency in the population relative to the non-resistant bacteria. If these are pathogens, you may inadvertently create a "superbug."

While mutations created the variation for selection to act up on, the selective force was the application of antibiotics, thereby allowing for the spread of resistance and the change in the population of phenotypes (to drug-resistant bacteria). Interestingly, not just antibiotics, but many other ingested substances (e.g., drugs and food) influence the gut microbiome and consequently our health. Microbes' role in metabolizing much of what is ingested, adds new genes, physiology, adaptation, and evolution for us to explore when considering the evolution of human health. The microbiome also challenges our understanding of the "environment" (since microbes are part of us but also separate in some ways – and mediate signals from the environment).

It is important to remember that evolution is not "purposeful" or directional, but rather is a product of differential reproductive success in specific environments. Like the evolution of antibiotic resistance, the evolution of extended lifespans and menopause may have important health implications and effects, but that does not mean that some force selected for or "designed" such traits.

Natural selection and artificial selection involve similar mechanisms but differ by whether humans are involved in the "selection" process (thus making it "artificial"). Darwin considered sexual selection to be an independent kind of selection altogether.

8.3.3 Sexual Selection

Colorful but cumbersome peacock tails are the classic trait highlighting the importance of sexual selection, an evolutionary mechanism by which individuals choose to mate with others with certain traits that appear to compromise survival but have reproductive benefits in certain contexts. While many evolutionary biologists see sexual selection as a special case of natural selection, Darwin himself saw it as a distinct and independent evolutionary mechanism. Within sexual selection, he described two potentially opposing mechanisms: same-sex competition (usually between males) for sexual access to or control over the opposite sex, and choice (often female) of mates based on aesthetic preferences.[11] Others have argued that seemingly non-adaptive traits like cumbersome peacock tails may signal better genes because signalers have survived costly burdens associated with "handicaps," thereby reframing sexual selection as a subset of natural selection for "handicaps."[12] People have critiqued the Handicap Hypothesis on theoretical and empirical grounds.[13] Interestingly, Prum suggested that patriarchal cultural resistance to the idea of female choice may have contributed to the belief that adaptation by natural selection is a universally strong force (adaptationism) and a tendency to want to subsume sexual selection under natural selection, so as to be able to explain evolution without invoking female choice.[14]

This bias against considering sexual selection as a separate mechanism of evolution has implications for our analysis of menopause and may explain why sexual selection is rarely considered, and why so many have searched for natural selection explanations for the existence of menopause. Mathematical modeling supports the possibility of sexual selection resulting in coevolution of mating preferences and displays that lead to a runaway process with traits that may negatively impact survival, but benefit reproductive success.[14] These models suggest potential tension between natural and sexual selection, and that these processes are best considered separately.

How we view sexual selection vis-à-vis natural selection is important for our understanding of the concept of fitness. In biology, the phrase "survival of the fittest" does not mean "fitness" in the sense of being able to lift weights or do aerobics at the gym. Rather, fitness is a measure of an individual's reproductive success (how many offspring one has). The use of the term "survival" is somewhat misleading here because survival without reproduction does not count in the game of evolution. The currency of evolution is genes passed to the next generation, so "reproduction of the fittest" and how many offspring you have are what count (from a purely biological perspective). Of course, in many contemporary human societies, people are rarely valued solely based on their reproductive success. Our cultural definitions of success include social, psychological, and economic perspectives. But from a strictly evolutionary perspective, fitness is a measure both of how well an organism is adapted to its environment *and* its success in mating and rearing offspring. In the evolutionary "game," only a slight advantage is needed to increase relative fitness of individuals with a particular trait in a population.

For Darwin, fitness meant the physical capacity to do tasks required for survival and reproduction. Later, population geneticists redefined fitness mathematically as the differential success of one's genes in subsequent generations, thereby combining survival, fecundity, and mating success into a single variable under the common label "adaptive natural selection." In so doing, the concept of fitness became mathematically convenient but rendered it difficult for people to think about distinct, independent mechanisms like sexual selection. By combining survival, fecundity, and mating into one variable, we may hinder our ability to consider hypotheses where these concepts conflict. This concept of "adaptive natural selection" is particularly problematic when looking at traits that may not seem to make adaptive sense, such as menopause. For a holistic evolutionary understanding of aging and menopause, we must consider the potential role of sexual selection as a process separate from natural selection.

8.4 Life History Theory

Just as there might be tensions and tradeoffs between sexual and natural selection, there are tradeoffs between how an organism invests in reproduction vs. growth and maintenance. Life History Theory (LHT) is a branch of evolutionary theory that examines the timing of life stages such as infancy, childhood, adulthood, and elderhood in terms of tradeoffs between growth and maintenance vs. reproduction, and tradeoffs in time and energy investments in current vs. future offspring, from a comparative perspective.[15] These issues are highly relevant to a discussion of aging and menopause, and a post-reproductive senescent life stage, as well as the decoupling of processes involved in aging of the reproductive system and the rest of the body (somatic).

Similar to anthropologists comparing cultures to understand patterns of behavior in humans and to assess how environments shape culture and vice versa (see Part II), evolutionary biologists use Life History Theory to compare species in different environments and predict: when organisms should begin reproducing, how many offspring they should produce, how much they should invest in those offspring vs. producing new offspring, and when they should cease reproducing.[15] Life History Theory involves fitness tradeoffs which vary in different environments. Generally, species that have high adult mortality in their environment do better (in terms of fitness) if they grow quickly and reproduce early, as opposed to growing slowly and reproducing late. Species with high juvenile mortality do better (in terms of fitness) if they produce many offspring and invest minimally, compared to producing few offspring with great parental investment. Once organisms reach maturity, Life History Theory suggests that they should reproduce until the fitness costs of another pregnancy (including costs to existing children, increased risk of maternal mortality, and increased risks of genetic abnormalities) are greater than the potential benefit of having another child. Human child dependence on mothers and others, including grandmothers, not only for food, but also other physical and cultural resources, makes calculating these tradeoffs difficult.

8.4.1 Life History across the Human Life Course

How is Life History Theory relevant for humans? What can we learn from comparisons to other mammals and primates? How do humans depart from other animals in terms of life history? Humans show remarkable behavioral flexibility, variation, and adaptability to many environments. Life history tradeoffs influence the relative length of life cycle stages, which for humans follow a general pattern,[16] with ages influenced by biocultural historical and environmental factors (Table 8.1).

Compared to most other mammals, humans have relatively long juvenile stages, and especially long senescent or post-reproductive stages (particularly in females). Could these two longer stages be related? Proponents of the Grandmother Hypothesis (see Chapter 10) say yes, arguing that children's need for more adult investment may have selected for grandmothers to live beyond reproduction.

TABLE 8.1 Life history stages of humans, approximate timing, and key characteristics. Modified from Bogin[16]

Life history stage	*Approximate Timing*	*Key Characteristics*
Gestation	40 weeks before birth-birth	Conception to birth
Infancy	0–3 years	Rapid growth from birth through weaning
Childhood	3–7 years	Moderate growth and dental maturation
Juvenility	7–10/12 years (girls/boys)	Slower growth rate, self-sufficiency in feeding, cognitive transition
Adolescence	10/12 years–adulthood	Final growth spurt until adulthood (often occurs when sexual maturity is reached, but variable depending on biological and cultural factors [e.g., age 18 vs. 20 depending on legal and cultural criteria in Japan])
Adulthood	Sexual maturity until elderhood	Characterized by relative stability in physiology, behavior, and cognition until end of childbearing years
Elderhood	Variable	Decline in function and repair ability of physiological systems, including reproduction. Depending on the organ and the individual, functional decline may start as early as late teens, in 30s (female reproduction), 40s (eyesight), or 50s or beyond

In some contemporary human environments where knowledge, education, economic, and other resources may be deemed critical to long-term reproductive success, people often choose to delay reproduction (by delaying marriage or using contraception, when available). Humans may also modify life history stages by dietary, physical activity, and medical interventions (e.g., exercise can suppress ovulation and delay reproduction; dietary antioxidants and caloric restriction may increase life expectancy and extend senescent stage), as well as by cultural rules (about marriage and reproduction). In resource-constrained environments, reproduction may be suppressed, perhaps allowing increased investment in growth and maintenance, thereby permitting reproduction when environmental conditions are better. Thus, human life history is influenced by both biological and cultural factors.

8.4.2 Life History across Human History

The extent to which cultural shifts such as changes in agriculture, industrialization, reproduction, and fertility-related behaviors can impact life history tradeoffs highlight the importance of employing a biocultural approach in order to gain a holistic understanding of aging. Several transitions or "revolutions" in human history have impacted human life history patterns,[17] life expectancy, population structure, and the experience of aging. These include: (1) ancient demographic shift from ape to human; (2) Neolithic revolution which significantly modified diet, activity patterns, and environmental exposures with significant impacts on health and longevity; (3) Industrial revolution which significantly decreased infant mortality as well as later life mortality; (4) Fertility revolution, or Demographic Transition, characterized by a delay in first pregnancy, and fewer children in the average woman's life. Some of the characteristics of these transitions or revolutions are summarized in Table 8.2, although these characteristics may vary by population.

How did environmental changes select for traits conferring longevity, and relatedly, menopause? And when? Since the middle of the 20th century, commonly referred to as the Anthropocene, human modifications of the environment have been impacting human health and life expectancy throughout the world, thus demonstrating the capacity of humans to change their environments in ways that impact life history. These modifications include processed foods, industrial agriculture, use of antibiotics and other chemicals in food production, climate change, and zoonotic (animal-based) pandemics. Such changes, and associated behaviors, likely influence the experience of aging and menopause.

Evolutionary hypotheses for aging and menopause often hinge on the concept of adaptation (see Section 8.5), but we need to be careful to identify what the evolutionary biologist Stephen J Gould calls "Just So Stories" in evolutionary theory. These are stories about why a trait might be adaptive, even with little evidence that alternative variation existed for natural selection to act upon. Evolutionary hypotheses are often hard to test because we cannot easily go back in time and run experiments – especially on humans. So instead, we use a comparative approach. With traits that leave evidence in the fossil records (like bipedalism, skull size, and even longevity), we can compare organisms through time. But with traits that do not leave fossil evidence, like menopause, we need to compare related species, and even within species, to examine population differences and get clues as to how much a trait may be under biological control, how much variation may have existed for natural section to act upon, and what that variation looks like. When related species like bonobos are examined, many of our dearly held evolutionary "Just So Stories" of human evolution and human uniqueness are challenged.[18] Questions we need to ask include: What is the trait we are

TABLE 8.2 Historical human population transitions or revolutions and key characteristics

Transitions/ Revolutions	*Time frame*	*Hallmark changes*	*Life history changes*	*Environmental changes*	*Anatomical/ physiological changes*	*Behavioral changes*
Ancient/ evolutionary	6–7 million years ago	Separation from great ape-human ancestor	Lifespan doubled (compared to common ancestor), but higher infant and maternal mortality	Increased meat eating, fire use	Decreased body size, increased brain size, bipedalism, riskier birth	Evolution of language, alloparenting, fire, and cooking
Neolithic	12,000 years ago	Origin of agriculture, change in reliability and diversity of food supply	Decreased interbirth interval, increased child survival	Dramatic shifts in diet (increased grains, dairy in some populations) and activity levels	Increased dental caries, microbiome changes	Decreased foraging and increased farming; increased social stratification and inequity
Industrial	Mid-1700s to 1800s	Use of fossil fuels and machines	Decreased infant mortality and later age mortality	Decreased infectious load, increased access to energy-rich foods	Microbiome changes (due to diet and time spent indoors)	More time indoors, increased social inequity
Fertility	Late 1800s–present	Increased fecundity but decreased fertility	Decreased infant and child mortality, increased fecundity through potential shortening of interbirth interval, but decreased cultural and behavioral fertility in many populations	Increased hygiene and food; decreased fertility through marriage and contraception and medical interventions	Changes in lifetime hormonal exposures, microbiome changes (related to birth practices, also increased use of antibiotics)	Changes in women's roles

Modified from Finch.[17]

trying to explain? What was the ancestral trait, and what variation existed? In what historical environmental contexts could it have been adaptive and how can we assess that?

8.5 Adaptation

How do we know if a particular trait, such as long lifespan or menopause, is adaptive? An adaptation is a trait that gives an organism a survival or reproductive advantage in a particular environment, although not necessarily in their current environment, or at all times in the life course. For example, bipedalism (walking on two legs instead of four) appears to be an adaptive trait that freed up hands for gathering food, using tools and carrying infants, and occurred around the time of our ancestral split from the common ancestor of chimpanzees and our lineage. Big brains evolved later and appear to be adaptive, providing our lineage with abilities that have allowed us to survive and reproduce in most terrestrial habitats on the planet. Yet, these two adaptive traits also entail significant evolutionary tradeoffs that occur when selection cannot change one part of a biological system to increase fitness without incurring fitness costs due to another system. Big brains and changes in the pelvis to allow bipedalism made human birth very difficult – with a convoluted fetal pathway through the birth canal, which increases infant and maternal mortality, and thus has significant impacts on reproductive success.[19]

Despite these increases in survival and reproduction risks, large brains must have had adaptive benefits that outweighed these real costs. And shifts in life history that allowed human babies to be born at a less developed stage, with prolonged dependence after birth, were one "solution" to the tradeoffs between the anatomically large brains and bipedalism-associated pelvic changes.[19] Of course, bipedalism has also led to later life health costs, such as common lower back pain and knee issues. These are unintended later life consequences resulting from tradeoffs and selective pressures acting earlier in the life course, but occurring after reproduction is mostly completed and when selection is weak.

There are several reasons that evolution may not lead to optimally adapted organisms in a particular environment. Anthropomorphically, evolution "tinkers" with existing structures and genes, but is not able to invent traits *de novo* or design them optimally. Lack of variation, developmental constraints, and changes in environment (with inadequate time or genetic variation to allow adaptation) are alternative explanations to natural selection that must be considered when evaluating evolutionary hypotheses about a particular trait, such as aging or menopause. For example, perhaps the environment has changed so that being post-reproductive now has health costs like cardiovascular disease or reproductive cancers that were not issues for our ancestors. We must ask whether a trait: is currently adaptive, was ever adaptive, or resulted from historical and developmental constraints and tradeoffs or sexual selection (meaning that it might compromise survival)? Work in evolutionary medicine has explored these alternative explanations for aging, which we discuss in Chapter 9.

8.5.1 *Understanding Selective Pressures and Tradeoffs: Skin Color and Skin Cancer*

Before we examine aging and menopause, it is useful to consider an example about which many misconceptions exist, despite the scientific evidence being relatively strong. Skin color is a trait that varies among humans due to varying levels of melanin pigment. Could skin cancer (melanoma) have been a primary factor in the evolution of skin color? Melanoma is diagnosed on average at age 66, and is 30 times more common among Whites than among Blacks,[20] suggesting to many people that the skin color trait (melanin) may affect differential survival. However, while early detection and treatments have improved survival rate, such interventions are recent on an evolutionary time scale. Given the risk of skin cancer associated with lighter skin, why did some humans evolve less melanin? In what environments might lighter skin be adaptive? What alternative explanations might there be? Phrased differently, if skin color is the trait we are trying to explain, could skin cancer be a relevant selective pressure? If so, then Whites, who are more susceptible to skin cancer particularly in higher latitudes, would appear less adapted to the environment than Blacks.

Skin cancer usually occurs after reproduction is completed and therefore does not affect reproductive success or fitness significantly. Selection is likely to be weak at that point in life, since an individual's genes have already gotten into the next generation. Even if skin cancer killed someone, genes for melanin that may predispose someone to skin cancer would already be in their descendants. Natural selection cannot act against genes that affect post-reproductive traits and survival rates. Do we need to look earlier in the life course to understand factors that may predispose certain individuals to later life health risks (e.g., Whites vs Blacks to skin cancer)?

The example of skin color is very informative and has been eloquently described by Jablonski.[21] As humans started migrating out of Africa about 100,000 years ago, skin color lightened over time – but why? In Part I, we discussed the importance of calcium for bone growth. Calcium absorption requires vitamin D, which in turn requires sunlight for synthesis. Without adequate vitamin D, a condition called rickets may occur. This is characterized by compromised calcification, softening, and distortion of bones, resulting in bowing of legs and compacted pelvis that may result in difficulties with childbirth and death of the baby and mother. Obviously, this is very important for reproductive success! In environments with less ultraviolet radiation (e.g., farther from equator), selection favors lighter skin since such individuals will be able to synthesize more vitamin D and lower their risk of rickets.

But why did ancestral humans have darker skin in the first place? Taking an evolutionary perspective, looking at human ancestors as well as our ape relatives, we observe that the skin of most apes is light-colored but covered with hair. As our ancestors living near the equator in Africa lost their hair, their light-colored skin would have led to breakdown of folate, which is fragile to UV radiation. With inadequate folate, neural tube defects like spina bifida and impaired sperm production often occur, and these compromise reproductive success. Among our ancestors, a mutation must have arisen that conferred greater melanin and darker skin, and this would have protected them from breakdown of folate, thereby conferring greater reproductive success, while hair might have been lost (for reasons not discussed here).

Reproduction plays a central role in determining how much melanin is adaptive in a particular environment, balancing the risk of neural tube defects and impaired sperm production (with excess UV exposure) with the risk of rickets (due to inadequate sunlight exposure and vitamin D production). Thus, there is an adaptive "sweet spot" for skin color in any given environment depending on the selective pressure of UV exposure. Recall that this is before sunscreen, UV-blocking fabrics, and people spending 90% of their time indoors. This example highlights how tradeoffs associated with a particular trait (e.g., skin color) may have advantages or disadvantages in different environmental contexts. Using rickets (due to lack of vitamin D) and spina bifida (due to lack of folate) as selective pressures on human evolution, we can explain the source of the human skin-color spectrum. Modern conveniences and technology allow humans to supplement food with vitamin D and folate, and put sunscreen and clothing on to block out UV radiation. Such inventions have allowed us to adapt to numerous environments, but the biology that underlies the skin color variation is still there, and it is important to understand why that variation exists, what its original function was evolutionarily, and its current health risks. By understanding the evolutionary biology, we may also avoid ascribing more significance to skin color than is biologically justified.

This case also highlights how health outcomes which usually occur post-reproduction (like skin cancer) are unlikely to exert significant selective forces – and highlights the importance of looking for selective forces earlier in the life course. Considering phenomena like aging- and postmenopausal-health issues, should we also examine potential roles for early life pressures and constraints?

8.5.2 Dangers of Misleading Biological Determinism

Not understanding the selective forces shaping skin color increases the likelihood of misunderstandings about health risks associated with skin color, misunderstandings that are often perpetuated through racialized medicine.

While skin color may be a proxy for ancestry and exposure to selective pressures that have shaped disease risk, it is critical to understand the nuanced ways in which such shorthand beliefs fall short. Biocultural examinations, such as those encouraged by the biocultural framework introduced in Chapter 1, challenge stereotypes that are grounded in oversimplified, misleading biological determinism.

Medical research institutions and researchers themselves often reinforce these stereotypes through methodological approaches and requirements. Since 1993, NIH has required researchers to include women and racialized minorities in their research populations, to ensure that they are being studied and perhaps even to expand understandings beyond the "universal white body" described in Part I. However, this requirement often serves to reinforce the idea that race and gender are real biological categories with neat divisions between them. Neither race nor gender are scientifically valid biological categories,[22] but rather they are social categories that can become embodied,[23] with significant biological impacts.

Whites are often the default in much medical research, and often non-Whites have poorer health. But in the case of osteoporosis, Whites seem to have worse health than Blacks in the United States. This may result from differences in biological causation, or there may be correlations with cultural factors influencing Blacks getting diagnosed less frequently due to discrimination and poverty which disproportionately affect Blacks compared to Whites due to structural racism (see Figure 2.2 on correlation vs. causation). Or more likely, there may be complex biocultural interactions, leading to population differences in local biologies. With all health conditions, it is important to be wary of biological determinism. Biological determinism occurs not only in the context of skin color and disease risk; it also occurs in the context of age and sex determinism about disease, often biased by reliance on Western subjects or patients, which constitute a narrow range of variation and are often assumed to be "normal" although they are WEIRD (Western Educated Industrialized Rich Democratic) (see Chapter 2).

8.5.3 Implications for Aging and Menopause

Part of anthropology's mission is to diversify perspectives and understandings by expanding data to include people of non-WEIRD populations, lifestyles, and environments. This enables us to assess what phenotypic variation exists among humans and better understand the factors that influence this variation in local biologies. The variation we observe is not solely due to genetic differences, but rather interactions between genes and environments that lead to phenotypic variation. Traits like skin cancer, and also aging and menopause, that occur later in life are not likely to result from later life selective pressures, and thus we may need to examine tradeoffs arising from selective pressures that act earlier in the life course.

The case study of evolution of skin color highlights early life tradeoffs (between vitamin D vs. folate) in specific environments with varying UV exposure and helps us understand why these are likely to have played a bigger role than skin cancer risk. Could menopausal symptoms or osteoporosis constitute significant selective pressures on the physiology of the female reproductive system, or are they like skin cancer – having impacts too late to really play a role? Should we be looking for evolutionary explanations earlier in the life course rather than focusing on the time around menopause? Can we think of the number of ovarian follicles and the hormones they produce as similar to the amount of melanin, with reproduction being compromised if hormones are too low or too high, and with selection acting to optimize fecundability (probability of becoming pregnant) in the 20s and 30s?

Explorations of selection, Life History Theory, and adaptation, with a consideration of skin color, laid the groundwork for considering evolutionary questions about why aging and menopause exist. Evolution generally prioritizes reproduction over the rest of the body. In an extreme view, the body only serves as a vessel to carry and spread genes, thus creating potential tradeoffs between investments in reproduction vs. maintenance of the body, with implications for aging and life expectancy. In Chapter 9, Ultimate Biology of Aging, we examine adaptive and non-adaptive hypotheses for aging such as "wear-and-tear" and the history of decoupling of somatic (non-reproductive body tissues) and reproductive (organs involved in procreation) biological aging (senescence). Then, in Chapter 10, Ultimate Biology of Menopause, we examine human post-reproductive lifespans compared to other species, and consider adaptive and non-adaptive hypotheses for menopause.

References

1. WHO. 2024. Life expectancy at birth. In: WHO, ed. Available at: https://data.who.int/indicators/i/A21CFC2/90E2E48
2. Dattani S, Rodés-Guirao L et al. 2023. Life expectancy. Available at: https://ourworldindata.org/life-expectancy. Accessed 23 Jan, 2025.
3. Dattani S. 2023. How does the risk of death change as we age – And how has this changed over time? Available at: https://ourworldindata.org/how-do-the-risks-of-death-change-as-people-age. Accessed 24 Jan, 2025.
4. Spadaro O, Youm Y et al. 2022. Caloric restriction in humans reveals immunometabolic regulators of health span. *Science.* 375(6581):671–677. https://doi.org/10.1126/science.abg7292
5. UC Museum of Paleontology. Evolution 101. *Understanding Evolution.* Available at: https://evolution.berkeley.edu/evolution-101/. Accessed 9 Feb 25, 2025.
6. Lind MI, Spagopoulou F. 2018. Evolutionary consequences of epigenetic inheritance. *Heredity.* 121(3):205–209. https://doi.org/10.1038/s41437-018-0113-y
7. Gilbert JA, Blaser MJ et al. 2018. Current understanding of the human microbiome. *Nat Med.* 24(4):392–400. https://doi.org/10.1038/nm.4517
8. Perez MF, Lehner B. 2019. Intergenerational and transgenerational epigenetic inheritance in animals. *Nat Cell Biol.* 21(2):143–151. https://doi.org/10.1038/s41556-018-0242-9

9. Henry LP, Bruijning M et al. 2021. The microbiome extends host evolutionary potential. *Nat Commun.* 12(1):5141. https://doi.org/10.1038/s41467-021-25315-x
10. Prestinaci F, Pezzotti P, Pantosti A. 2015. Antimicrobial resistance: A global multifaceted phenomenon. *Pathog Glob Health.* 109(7):309–318. https://doi.org/10.1179/2047773215y.0000000030
11. Darwin C. 1871. *The Descent of Man, and Selection in Relation to Sex*. London: J. Murray. Available at: https://www.loc.gov/item/04033382/
12. Zahavi A. 1975. Mate selection—A selection for a handicap. *J Theor Biol.* 53(1):205–214. https://doi.org/10.1016/0022-5193(75)90111-3
13. Penn DJ, Számadó S. 2020. The Handicap Principle: How an erroneous hypothesis became a scientific principle. *Biol Rev Camb Philos Soc.* 95(1):267–290. https://doi.org/10.1111/brv.12563
14. Prum RO. 2017. *The Evolution of Beauty: How Darwin's Forgotten Theory of Mate Choice Shapes the Animal World - And Us.* New York: Doubleday.
15. Stearns SC. 1992. *The Evolution of Life Histories.* New York: Oxford University Press.
16. Bogin B. 2015. Ch 20. Human Growth and Development. In Michael P Meuhlenbein (ed), *Basics in Human Evolution*, 285–293. Academic Press, London. https://doi.org/10.1016/B978-012156651-7/50015-0
17. Finch CE. 2007. *The Biology of Human Longevity: Inflammation, Nutrition and Aging in the Evolution of Lifespans.* Burlington MA: Academic Press.
18. Diogo R. 2018. First detailed anatomical study of bonobos reveals intra-specific variations and exposes just-so stories of human evolution, bipedalism, and tool use. *Front Ecol Evol.* 6:53. https://doi.org/10.3389/fevo.2018.00053
19. Rosenberg KR, Trevathan WR. 2021. The obstetrical dilemma revisited - Revisited. In Sallie Han, Cecília Tomori (eds.), *The Routledge Hadnbook of Anthropology and Reproduction.* London: Routledge; 425–440.
20. ACS. 2025. *Key Statistics for Melanoma Skin Cancer.* Available at: https://www.cancer.org/cancer/types/melanoma-skin-cancer/about/key-statistics.html. Accessed 24 Jan, 2025.
21. Jablonski NG. 2021. Skin color and race. *Am J Phys Anthropol.* 175(2):437–447. https://doi.org/10.1002/ajpa.24200
22. Yudell M, Roberts D et al. 2016. Taking race out of human genetics. *Science.* 351(6273):564–565. https://doi.org/10.1126/science.aac4951
23. Gravlee CC. 2009. How race becomes biology: Embodiment of social inequality. *Am J Phys Anthropol.* 139(1):47–57. https://doi.org/10.1002/ajpa.20983

9

ULTIMATE BIOLOGY OF AGING

If evolution entails selection for survival and reproductive success, why does aging ever occur? Wouldn't selection act to increase survival and lifespan to enable greater reproductive success? Scientists began grappling with these evolutionary questions in the mid-20th century. In 1957, Williams bemoaned the neglect of the study of aging by non-medical biologists, criticized the analogic thinking that compared decline of organisms to decline of mechanical devices, argued for the need to consider natural selection and evolution, and applied the concept of pleiotropy (a single gene having multiple effects)[1] to the evolution of senescence[2] (cellular changes that lead to age-related disease and death).[3]

In the game of evolution, it is always better to reproduce more (relative to competitors). That may not be true from the perspective of living standards, quality of life, societal expectations, or climate change. But in this chapter, we set aside cultural factors (examined further in Parts II and IV), to examine our evolutionary heritage that has selected for maximal fitness (reproductive success). The evolved predispositions we carry inside our genes and the life history tradeoffs that exist have implications for processes and experiences of aging and possible interventions.[4]

9.1 Traits vs. Tradeoffs

In applying an ultimate biological perspective to aging, what trait are we trying to explain? Are we trying to explain longevity within humans (with greater female longevity compared to males), or within primates (with humans living longer than non-human primates), or the aging rate of various organ systems, or the decoupling of somatic vs. reproductive aging, or even

DOI: 10.4324/9781003644521-12

the upper limit to lifespan (if there is one)? Is there a gene(s) for aging that might "program" the rate of aging, upon which selection can act?

The Programmed Aging Hypothesis proposes that biological aging is caused by mechanisms that provide an evolutionary fitness benefit. No single gene determining rates of aging has been found, but many genes likely underlie the life history tradeoffs between investing in maintenance and repair of somatic tissue vs. reproduction.[5]

In 1952, Medawar proposed that selection for living and reproducing longer should decline with age, starting at the time when an organism becomes able to reproduce.[6] Since risk of death from external causes (e.g., starvation, disease, and accidents) will eventually lead to death for all organisms in a population, it does not make "evolutionary sense" for organisms to invest energy into living and reproducing forever. Environmental mortality risks exert selective pressures and influence life history differences between organisms. Lifespan might be a by-product of selection acting to optimize reproductive success earlier in life, at the expense of somatic maintenance, involving inherent life history tradeoffs.[5] Thus, there may not be a single trait of aging to explain, but rather a lifetime of tradeoffs that lead to aging and eventual death.

Because resources such as time and energy are generally limited for most organisms, there are inherent tradeoffs when allocating resources to one trait or body tissue, generally at the expense of others. This limits the range of possible phenotypes and life history strategies, and influences rates of aging of various organ systems and organisms. Life history tradeoffs between somatic and reproductive investment are likely critical to aging, but developmental and physiological constraints may be equally important.[7] Much research has focused on energy tradeoffs between growth, reproduction, and longevity. But these may need to be reevaluated and expanded to consider tradeoffs in function because processes optimized for early life function may lead to problems later in life when selection is weak.[8]

9.2 Is an Elderhood (Senescent) Stage of Life Unique to Humans?

Is it possible for humans to live forever? To answer that question, a comparative approach, examining other animals, may be informative. Do any animals live forever? If there are no genes that confer immortality, then natural selection cannot act on them. Living forever means being able to regenerate any body part no matter what kind of damage occurs. Stem cells allow for this, but generally, in complex organisms, stem cells differentiate and become different cell types (e.g., neurons, heart cells, muscle cells). Stem cells also run the risk of becoming cancerous, which is one way to gain immortality, as happened to Henrietta Lacks through her *HeLa* cancer cells, which are used throughout the world for research and were taken without her consent or compensation.[9] Stem cells and cancer cells are like two sides of the same coin, presenting an inherent tradeoff. On one side, stem cells can undergo many cell divisions to repair tissue (and thus promote survival) early in life, but, on the other side, in the process of division, they may accumulate mutations that can increase risk of uncontrolled growth (i.e., cancer), particularly later in life.

There is one animal that has come close to immortality. The hydra (related to corals, sea anemones, and jellyfish), measures only millimeters long and has regenerative capabilities that has earned it the name "eternal embryo" or "immortal hydra." This non-senescence (lack of biological aging) appears possible only in simple organisms like hydra which have high proportions and numbers of stem cells, continuous cell divisions, and an effective cell selection mechanism.[10] So, immortality is not impossible in the animal kingdom – but are you willing to become a hydra to gain immortality? I am not sure the tradeoff seems worth it....

Examples of animals with minimal biological aging and deterioration (senescence), such as naked mole rats, or ones with accelerated aging such as killifish, provide examples of the range of variation in aging processes that exist, and clues to drivers of aging as well as possible interventions.[11] Naked mole rats' mortality risk does not increase with age, in part due to antioxidant mechanisms, pointing to one potential proximate approach to slowing aging.[11] Killifish, which age quickly, can have their lifespan extended by recolonizing their guts with bacteria from younger fish.[12] This highlights an approach of introducing microbes (probiotics) and/or diets that support healthy microbes (prebiotics) associated with youth or health as a potential anti-aging intervention.

9.3 Evolutionary Theories of Aging

Ultimate biological, or evolutionary, theories explore why we age, why different species age at different rates, and why some individuals stop reproducing. There are many evolutionary theories for aging, and they are increasingly being integrated with new understandings about epigenetics,[13] symbioses and host-microbiome interactions,[14] and regulatory mechanisms at multiple scales.[15] Below we focus on a high-level review of some of these theories, examining non-adaptive and adaptive explanations for aging.

The Programmed Aging Hypothesis, often explained in terms of group selection, argues that aging and dying make room for younger individuals, and gains support from the over 200-fold difference in longevity across species of mammals.[16] While variation in aging rates clearly exist, this does not mean that aging rates have been selected for as an independent trait. Due to weaker selection later in life, direct selection for increased (programmed) aging later in life seems unlikely. Other theories, including the non-adaptive Mutation Accumulation[6] and the Wear-And-Tear theories, and the adaptive Antagonistic Pleiotropy[3] and Disposable Soma[5] theories, view aging as a side effect resulting from environmental interactions and tradeoffs across the life course.

9.3.1 Non-Adaptive Hypotheses for Aging

In 1952, Peter Medawar proposed the Mutation Accumulation Theory of Aging, arguing that beyond a certain age, the evolutionary benefit of a longer lifespan in mammals is negligible, and this "certain age" is related to the age at which reproduction becomes possible.[6] The Theory posits that all organisms accumulate effects of mutations that may occur every time their cells divide. While most mutations may not cause problems, over time, mutations accumulate and some cause problems.

Mutations are a proximate biological explanation for *how* aging happens (see Part I), while Mutation Accumulation Theory is an ultimate biological theory explaining *why* we age. Mutation rates vary by species (and even among males and females in mammals and birds)[17] and likely contribute to varying lifespans in different organisms.

As anyone who has experienced injuries recognizes, being alive has inherent risks. These can range from major accidents to minor daily stressors, and even to invisible, molecular changes that contribute to aging and mortality risk. Mutation rates can be increased by exposure to radiation and oxidative damage, including mitochondrial activity. Epigenetic changes to the genome influenced by environmental factors such as diet[18] accrue as one ages, and changes in the gut microbiome modulate aging,[19] so that functional declines associated with aging result from multiple processes that accumulate as by-products of simply living.

The Wear-And-Tear Theory[20] hinges on life history tradeoffs between reproduction and repair. In resource-constrained settings, once organisms begin reproducing, they may invest less in repair. Aging then occurs as a by-product or constraint of lifetime exposures. This implies that additional energy may be able to offset or overcome these constraints, allowing organisms to repair tissues while reproducing. Thus, one might expect wear-and-tear processes to be offset in some circumstances, when additional resources can be invested in repair.

9.3.2 Adaptive Tradeoffs

Another evolutionary approach considers aging as resulting from processes that are by-products of selection during the reproductive phase of life, involving potential tradeoffs between somatic and reproductive cells, and between earlier and later life health. Once organisms have reproduced and gotten their genes into the next generation, selection cannot act against those individuals who might be carrying genes that shorten life expectancy or increase risk of diseases associated with aging. And if any of their genes contribute to greater fitness, they will comprise a greater proportion of subsequent generations.

Kirkwood proposed the Disposable Soma Theory of Aging,[21] in which the body (soma) is conceptualized as a "disposable" container that may get harmed in the process of transporting and spreading genes from eggs and sperm into the next generation (Figure 9.1, panel A). In contrast to most vertebrate males that can generate sperm throughout the lifespan, most female birds and mammals lack ovarian stem cells in adulthood, and so cannot produce new gametes. (But see recent evidence of ovarian stem cells in birds.)[23] This leads to decoupling of reproductive and somatic aging. The pattern in female birds and mammals involves producing a large supply of eggs and the follicles which contain them. Over the life course these mature, and either get fertilized or degenerate, with the follicular cells producing the hormones needed for ovarian cycles. If organisms live beyond the time when follicle

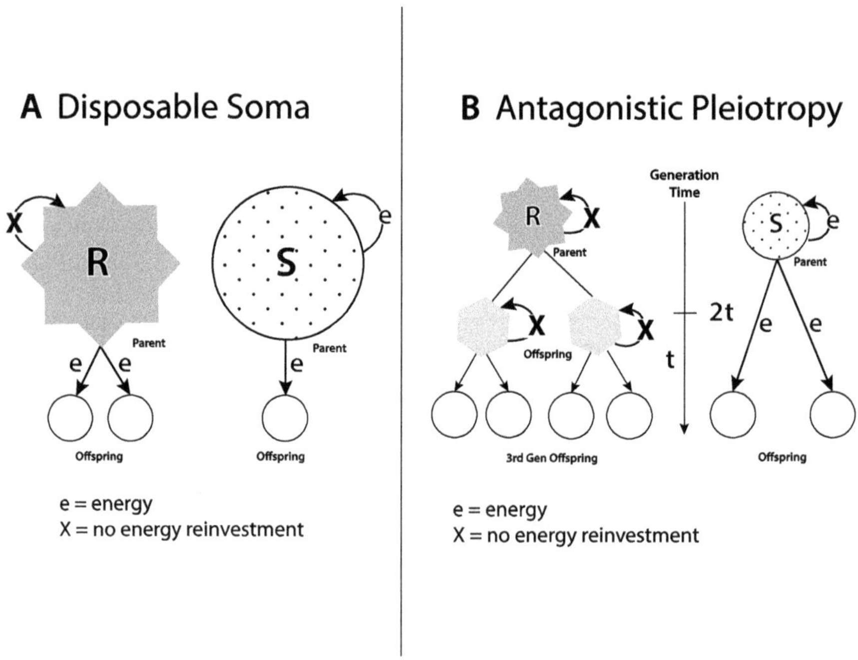

FIGURE 9.1 (A) Disposable soma and (B) antagonistic pleiotropy theories of aging. Modified from Trindade.[22]

numbers have fallen below the threshold required for hormone production, they become post-reproductive.

Figure 9.1 Panel A illustrates the Disposable Soma Theory: selection for longevity is countered by selection for prioritization of energy (e) used for reproduction. An individual prioritizing reproduction (R) uses energy to produce offspring, in contrast to an individual that uses more energy to maintain soma (S). Thus, R ages (declines) faster but produces more offspring than S, leading to greater representation of R's genes in subsequent generations.

Figure 9.1 Panel B illustrates the Antagonistic Pleiotropy Theory: selection for longevity is countered by selection for faster reproduction. An individual with genes conferring reproductive advantage (R) but negative impact on body homeostasis or maintenance has a shorter generation time (t) to produce two offspring. Compared to another individual who reproduces more slowly and invests more resources in somatic maintenance (S), we observe that after 2t, individual R has four grand offspring compared to individual S's two offspring. Thus, while individual S may live longer, individual R has higher fitness.

Organisms that appear to invest in soma instead of reproduction later in life seem to present an evolutionary puzzle. This is why the existence of menopause has intrigued so many. Reproduction shuts down, but the body keeps going. Menopause entails what seems to be premature aging of reproductive cells before the somatic cells, or somatic cells that somehow outlive their "usefulness" for reproduction. However, in organisms with high parental investment, maintenance of soma may be necessary for successfully rearing offspring to adulthood, thus ensuring long-term reproductive fitness. Also, in species lacking ovarian stem cells in adulthood, it is not clear how additional investment in reproduction could lead to increased fitness due to the evolutionary and developmental constraints of the system (i.e., since the number of eggs or hormones produced by surrounding cells of follicles can't be easily changed).

Life History Theory hypothesizes that dangerous environments shift reproduction earlier and are associated with shorter lifespans. In contrast, safer environments allow for slowing of life history, with later reproduction and longer lifespans. To what extent have humans, compared to non-human primates, shaped and modified their environments to make them safer, through use of fire, and decreased reactive aggression (or self-domestication)?[24] Could such changes have led to longer human lifespans?

The Antagonistic Pleiotropy Theory of aging proposes that aging occurs as an unintended by-product of optimizing adaptive mechanisms influencing earlier growth, survival, and reproduction (Figure 9.1, panel B). Pleiotropy refers to genes having effects on different traits. Antagonistic pleiotropy entails contradictory or opposing effects of a single gene during different life stages. This theory posits that changes (including interventions by humans) to genes or physiological processes that might extend life expectancy might potentially sacrifice fitness by imposing constraints on early life.

Examples of genes that exhibit antagonistic pleiotropy and confer advantages to survival and reproduction earlier in life but may shorten lifespan by causing health problems later in life when selection is weaker include:

1 Genes that appear to be advantageous for reproduction early in life but contribute to increased risk of coronary artery disease.[25] If individuals with these genes reproduce more than others, then individuals carrying such genes will increase in the population, contributing to increases in coronary artery disease.
2 Genes involved in immune function that protect against infection early in life may contribute to brain inflammation and neurodegenerative disease later in life.[26] While nobody wants to have neurodegenerative diseases later in life, it may be a tradeoff for surviving infection early in life.

Recall how melanin impacts later life skin cancer risk but derives from adaptive tradeoffs between earlier life health risks associated with folate and vitamin D synthesis in environments with varying UV exposure (see Chapter 8). In the context of menopause, we often focus on later life hormones, but could hormones influencing menopausal symptoms and health risks be influenced by selective factors acting on earlier life health? If so, should we exercise caution about when and how we intervene, for example with hormonal birth control? Or is it possible to intervene in such a way that we can have our cake (in early life) and eat it too (in later life)? Understanding the evolution of traits related to health, and potential tradeoffs involved, can help us understand if, when, and how we might intervene. These are the very questions that have led to the field of evolutionary medicine.

9.4 Evolutionary Medicine

Why is it important to understand whether a symptom or trait is adaptive or not? If a symptom is caused by a pathogen or some bodily dysfunction, then treating it makes sense. However, if it is adaptive, and has a function, then it might be better not to treat it, or to treat it differently. For example, while conventional medical advice has been to treat fever with medicines that reduce fever (antipyretics) in order to minimize discomfort and risk of seizures, by doing so, we appear to be interfering with an evolved defense mechanism, thereby impairing our body's ability to protect itself, and actually increasing risk of infection.[27] Thus, when considering health-related symptoms or traits, it is important to have an understanding of potentially adaptive functions or alternative explanations before we decide how best to treat, if at all.

However, not all health conditions are as adaptive as fever. The field of evolutionary medicine outlines six evolutionary reasons besides selection for why symptoms or disease might appear:[28]

1 Mismatch: our bodies are unprepared to cope with modern environments that have changed dramatically since our paleolithic (hunter-gatherer) existence (e.g., obesity caused by modern obesogenic environments).
2 Infection: bacteria and viruses evolve faster than we do (e.g., why antibiotic resistance appears almost as soon as we start using new antibiotics).
3 Constraints: evolutionary (phylogenetic) history or developmental constraints (physical, mechanical, or structural limitations as well as irreversible commitments at key developmental stages) may limit variation for natural selection to act upon (e.g., choking is possible because of anatomical constraints of relative locations of the "pipes" to lungs and stomach).
4 Tradeoffs: between somatic vs. reproductive cells, and genes acting at early vs. later life history stages (e.g., antagonistic pleiotropy).
5 Reproduction: natural selection maximizes reproduction, not health; sexual selection may lead to traits that increase reproductive success but lead to decreased later life health and/or increased mortality risks.
6 Defensive responses: like fever, responses such as pain and inflammation may be useful in certain environments but may cause unwanted discomfort and aging-related symptoms that we want to medicate.

Considering aging and post-reproductive health for Japanese women from an evolutionary perspective, several of the six reasons listed above may play a role. While genes most likely play a role in some health differentials and longevity,[29] data on Japanese women living in the United States vs. Japan suggest that at least for cardiovascular disease[30] and menopausal symptoms such as hot flushes,[31] genetics play a relatively minor role compared to environmental factors, as symptoms are greater among Japanese living in the United States than those living in Japan. This is promising in terms of the potential for interventions to reduce later life disease and symptoms.

Many evolutionary approaches to health-related phenomena have attributed disease to a mismatch between the pre-agricultural environment in which our ancestors lived for hundreds of thousands of years, and contemporary WEIRD environments and lifestyles. Such mismatches may account for the rise in reproductive cancers, as well as in symptoms associated with menopause. Yet Japanese women, despite having later age at first birth, fewer children, and shorter breastfeeding duration than many other, industrialized countries, nevertheless have lower rates of reproductive cancers and menopausal symptoms and are thus more typical of natural fertility populations. Which aspects of their environment buffer them from the deleterious effects of post-industrial lifestyles?

One potential environmental factor in Japan that may influence the experience of aging, menopause, and postmenopausal health is the relatively high intake of soy, which contains plant estrogens (phytoestrogens). These phytoestrogens, and their metabolites (which are influenced by the microbiome composition of an individual's gut), bind to estrogen receptors, and thus have the potential to exert effects on reproductive endocrinology and modulate the experience of menopause. Thus, culture, as expressed through diet, interacting with biology, in the form of microbiome and soy phytoestrogens, may have significant physiological, and possibly epigenetic effects, on health, aging, and menopause.

Soy phytoestrogen modulation of reproductive ecology (mediated by the microbiome) might provide one alternative path to address the mismatch between evolutionary and current lifestyles. While theoretically interesting, it also has implications for how we think of and deal with menopause.

9.5 Is Dairy Intake Necessary to Prevent Osteoporosis?: An Evolutionary Example

As explored at the close of Part I, osteoporosis is a commonly reported disease of aging, characterized by thinning of bone tissues through decreased deposition and increased uptake, leading to bone density loss and increased fracture risk. Increased calcium intake, particularly from dairy products, and weight-bearing exercises, are common recommendations for preventing osteoporosis. Ultimate biological approaches urge us to question this conventional wisdom to better understand relationships between genes and environmental factors and health outcomes such as osteoporosis.

Is osteoporosis an inevitable part of aging? A comparative perspective reveals that osteoporosis prevalence rates vary between populations and within populations by sex and ethnicity. Women have higher rates of osteoporosis compared to men, and non-Hispanic White women in the United States have higher rates than other groups.[32]

Is calcium from dairy required to decrease osteoporosis risk? From an evolutionary perspective, this seems puzzling because dairy is a relatively recent addition to the human diet, so how could it be required to prevent osteoporosis? All mammals produce the enzyme lactase to digest lactose in their mothers' milk. But after the age of weaning, all other mammals, and most humans, lose the ability to digest lactose. Western biomedicine labels people as either being lactose tolerant (able to digest lactose, because they have the enzyme lactase) or lactose intolerant (unable to digest lactose, because they no longer produce lactase). But from an evolutionary and comparative perspective, the ability to produce lactase enzyme into adulthood is completely novel and would be wasteful since for most mammals there is no dietary source of lactose once a baby is weaned from its mother. The ability to produce lactase enzyme past the age of weaning is not even that widespread among humans, but mutations conferring persistence of lactase enzyme arose in several geographic locations (including the Middle East, Western Africa, and Northern Europe), originating in populations with a history of dairying.

Interestingly, there is also a "calcium paradox:" fracture rates are the highest in countries with the highest calcium and dairy product intakes, and the lowest in countries with low intakes of dairy.[33,34] If calcium is essential for strong bones and prevention of osteoporosis, wouldn't we expect the opposite relationship? Perhaps there are confounding factors that play a role. Interestingly, despite not consuming dairy, some contemporary hunter-gatherers have twice as much dietary calcium intake compared to industrialized populations, with their calcium likely coming from plants and fish.[35] But how might high calcium intake be correlated with high fracture rates?

To minimize the influence of potentially confounding variables, researchers try to examine populations that share many variables, but may differ primarily in only the variables of interest. Hilliard has examined two groups of African women with different rates of dairy intake and osteoporosis: West African women with lower rates of osteoporosis hip fractures compared to East African women, with whom they share many similarities in terms of genes.[36] One big difference is their dairy farming practices. The West Africans live in regions with tsetse flies that attack cattle, thus preventing dairy farming; the East Africans are cattle-farmers. Without cattle, the West African populations would have had no selection for lactase persistence. The East African populations have higher lactase persistence, enabling lifelong consumption of dairy, but also seem to have increased risk of osteoporosis. How can this be?

Hilliard proposes an intriguing evolutionary hypothesis, that selection for lactase persistence in the East African populations may have also selected for some level of need for dairy due to changes in calcium homeostasis.[36] That is, in expanding their dietary repertoire and acquiring the ability to digest lactose later in life, they may have inadvertently increased their dependence on dairy, and their risk of osteoporosis. Since fractures resulting from osteoporosis tend to affect older individuals later in life, selection against mutations that increase osteoporosis risk would be weak, especially if the mutations have adaptive benefits early in life by allowing carriers of such genes to exploit a novel food source of other mammals' milk. Such mutations may have been instrumental to the reproductive success of individuals in these dairy-producing populations.

While speculative, this evolutionary hypothesis raises questions about the consequences of recommending dairy products to lactase non-persistent populations (who may exhibit bone health and low osteoporosis risk without dairy consumption). It also questions one size fits all recommendations for any health condition, especially ones that emerge later in life such as postmenopausal health issues, when selection is weak, that may be influenced by a lifetime of factors. In Chapter 10, we will apply similar evolutionary thinking to the phenomena of menopause and post-reproductive lifespans.

References

1. Stearns FW. 2010. One hundred years of pleiotropy: A retrospective. *Genetics.* 186(3):767–773. https://doi.org/10.1534/genetics.110.122549
2. McHugh D, Gil J. 2018. Senescence and aging: Causes, consequences, and therapeutic avenues. *J Cell Biol.* 217(1):65–77. https://doi.org/10.1083/jcb.201708092
3. Williams GC. 2001. Pleiotropy, natural selection, and the evolution of senescence. *Sci Aging Knowl Environ.* 2001(1):cp13–cp13. https://doi.org/10.1126/sageke.2001.1.cp13
4. Flatt T, Partridge L. 2018. Horizons in the evolution of aging. *BMC Biol.* 16(1):93. https://doi.org/10.1186/s12915-018-0562-z
5. Kirkwood TBL. 2005. Understanding the odd science of aging. *Cell.* 120(4):437–447. https://doi.org/10.1016/j.cell.2005.01.027
6. Medawar PB. 1952. An Unsolved Problem of Biology*: An Inaugural Lecture Delivered at University College, London, 6 December, 1951.* United Kingdom: Taylor & Francis Group. Available at: https://books.google.com/books?id=ozEmPQAACAAJ
7. Cohen AA, Coste CFD et al. 2020. Are trade-offs really the key drivers of ageing and life span? *Funct Ecol.* 34(1):153–166. https://doi.org/10.1111/1365-2435.13444
8. Maklakov AA, Chapman T. 2019. Evolution of ageing as a tangle of trade-offs: Energy versus function. *Proc Biol Sci.* 286(1911):20191604. https://doi.org/10.1098/rspb.2019.1604
9. Skloot R. 2010. *The Immortal Life of Henrietta Lacks.* New York: Crown Publishers.
10. Dańko MJ, Kozłowski J, Schaible R. 2015. Unraveling the non-senescence phenomenon in Hydra. *J Theoret Biol.* 382:137–149. https://doi.org/10.1016/j.jtbi.2015.06.043
11. Stenvinkel P, Shiels PG. 2019. Long-lived animals with negligible senescence: Clues for ageing research. *Biochem Soc Trans.* 47(4):1157–1164. https://doi.org/10.1042/bst20190105
12. Smith P, Willemsen D et al. 2017. Regulation of life span by the gut microbiota in the short-lived African turquoise killifish. *eLife.* 6:e27014. https://doi.org/10.7554/eLife.27014
13. McAuley MT. 2023. An evolutionary perspective of lifespan and epigenetic inheritance. *Exp Gerontol.* 179:112256. https://doi.org/10.1016/j.exger.2023.112256
14. Bapteste E, Huneman P et al. 2023. Expanding evolutionary theories of ageing to better account for symbioses and interactions throughout the Web of Life. *Ageing Res Rev.* 89:101982. https://doi.org/10.1016/j.arr.2023.101982
15. Cohen AA, Ferrucci L et al. 2022. A complex systems approach to aging biology. *Nat Aging.* 2(7):580–591. https://doi.org/10.1038/s43587-022-00252-6
16. Pamplona R, Jové M et al. 2023. Programmed versus non-programmed evolution of aging. What is the evidence? *Exp Gerontol.* 175:112162. https://doi.org/10.1016/j.exger.2023.112162
17. Bergeron LA, Besenbacher S et al. 2023. Evolution of the germline mutation rate across vertebrates. *Nature.* 615(7951):285–291. https://doi.org/10.1038/s41586-023-05752-y

18. Choi S-W, Friso S. 2010. Epigenetics: A new bridge between nutrition and health. *Adv Nutr.* 1(1):8–16. https://doi.org/10.3945/an.110.1004
19. Ghosh TS, Shanahan F, O'Toole PW. 2022. The gut microbiome as a modulator of healthy ageing. *Nat Rev Gastroenterol Hepatol.* 19(9):565–584. https://doi.org/10.1038/s41575-022-00605-x
20. Jin K. 2010. Modern biological theories of aging. *Aging Dis.* 1(2):72–74.
21. Kirkwood TL, Kapahi P, Shanley DP. 2000. Evolution, stress, and longevity. *J Anat.* 197(Pt4):587–590. https://doi.org/10.1046/j.1469-7580.2000.19740587.x
22. Trindade LS, Aigaki T, Peixoto AA, Balduino A, Manica da Cruz IB, Heddle JG. 2013. A novel classification system for evolutionary aging theories. *Front Genet.* 4: 1–8. https://doi.org/10.3389/fgene.2013.00025
23. Meng L, Zhang Y et al. 2023. Identification of oogonial stem cells in chicken ovary. *Cell Prolif.* 56(3):e13371. https://doi.org/10.1111/cpr.13371
24. Wrangham RW. 2019. Hypotheses for the evolution of reduced reactive aggression in the context of human self-domestication. *Front Psychol.* 10:1914. https://doi.org/10.3389/fpsyg.2019.01914
25. Byars SG, Huang QQ et al. 2017. Genetic loci associated with coronary artery disease harbor evidence of selection and antagonistic pleiotropy. *PLoS Genet.* 13(6):e1006328. https://doi.org/10.1371/journal.pgen.1006328
26. Provenzano F, Deleidi M. 2021. Reassessing neurodegenerative disease: Immune protection pathways and antagonistic pleiotropy. *Trends Neurosci.* 44(10): 771–780. https://doi.org/10.1016/j.tins.2021.06.006
27. Best EV, Schwartz MD. 2014. Fever. *Evolution, Medicine, and Public Health.* 2014(1):92–92. https://doi.org/10.1093/emph/eou014
28. Nesse RM, Williams GC. 1994. *Why We Get Sick : The New Science of Darwinian Medicine.* 1st ed. New York: Times Books.
29. Willcox DC, Willcox BJ et al. 2006. Genetic determinants of exceptional human longevity: Insights from the Okinawa Centenarian study. *Age (Dordr).* 28(4):313–332. https://doi.org/10.1007/s11357-006-9020-x
30. Marmot MG, Syme SL. 1976. Acculturation and coronary heart disease in Japanese-Americans. *Am J Epidemiol.* 104(3):225–247. https://doi.org/10.1093/oxfordjournals.aje.a112296
31. Gold EB, Block G et al. 2004. Lifestyle and demographic factors in relation to vasomotor symptoms: Baseline results from the study of women's health across the nation. *Am J Epidemiol.* 159(12):1189–1199. https://doi.org/10.1093/aje/kwh168
32. Noel SE, Santos MP, Wright NC. 2021. Racial and ethnic disparities in bone health and outcomes in the United States. *J Bone Miner Res.* 36(10):1881–1905. https://doi.org/10.1002/jbmr.4417
33. Frassetto LA, Todd KM et al. 2000. Worldwide incidence of hip fracture in elderly women: Relation to consumption of animal and vegetable foods. *J Gerontol A Biol Sci Med Sci.* 55(10):M585–592. https://doi.org/10.1093/gerona/55.10.m585
34. Hegsted DM. 2001. Fractures, calcium, and the modern diet. *Am J Clin Nutr.* 74(5):571–573. https://doi.org/10.1093/ajcn/74.5.571
35. Eaton SB, Konner M, Shostak M. 1988. Stone agers in the fast lane: Chronic degenerative diseases in evolutionary perspective. *Am J Med.* 84(4):739–749. https://doi.org/10.1016/0002-9343(88)90113-1
36. Hilliard CB. 2016. High osteoporosis risk among East Africans linked to lactase persistence genotype. *Bonekey Rep.* 5:803. https://doi.org/10.1038/bonekey.2016.30

10

ULTIMATE BIOLOGY OF MENOPAUSE

Why would menopause (the end of menstruation, signifying the end of reproductive potential) have originated? Why can human females spend a third of their adult lives not reproducing if reproductive success is so important? How can evolutionary perspectives (e.g., on adaptive function or constraints) challenge assumptions about what is "normal" (Part I), and influence expectations and interventions for menopause?

While adaptive hypotheses for the evolution of menopause, such as the grandmother hypothesis, have been proposed, evolutionary theory also includes non-adaptive explanations we should consider such as: weakening selection with increasing age, variation constrained by phylogenetic history and development, and mismatch between contemporary lifestyles and our evolutionary history. Also, we must recognize that selection acting on genes influencing aging will impact not only females but also males (since on average an individual will have roughly equal numbers of male and female offspring). Thus, to fully understand the evolution of menopause and begin to answer the above questions, we may need to consider multiple evolutionary factors and explanations.

In evolutionary thinking, we often draw on architectural analogies, particularly when examining anatomical structures, but such analogies may also be insightful when examining life history traits like menopause and the post-reproductive lifespan. In European churches, arches often support domes, and at the junction of two arches in the triangular spaces called spandrels, ornate decoration can often be found (Figure 10.1). In a classic paper, biologists Gould and Lewontin questioned the *purpose* of the spandrels of San Marco cathedral: "people who come into San Marco often remark that the architect was clever to design the church with those spaces to be decorated,

DOI: 10.4324/9781003644521-13

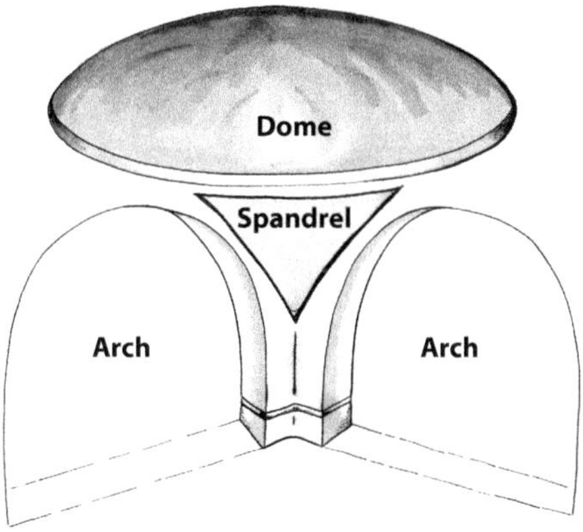

FIGURE 10.1 Architectural sketch of supported dome, identifying the arches, and the spandrel that is created by the arches (Drawing by kristen melby). This architectural image provides a model for menopause as a figurative spandrel, incidental to other forces illustrated by the figurative arches of reproductive and somatic function and aging.

but it turns out that that is not why they are there...."[1] Gould and Lewontin explain that the decorations were incidental to the more practical function of supporting the dome. They use the metaphor of spandrels to question "adaptationist" thinking, common in evolutionary biology, that assumes that traits that seem to provide fitness benefits must have been selected for by natural selection and be adaptive. Just as with highly decorated spandrels, we need to recognize that how a trait currently functions may differ considerably from how it originated.

Similarly, in the case of menopause, people routinely assume that there must be fitness benefits to menopause, arguing that otherwise it would not exist. But in doing so, they are ignoring constraints. Many researchers have spent countless hours generating and collecting data to support possible "Just So Stories" (see Chapter 8) and hypotheses that suggest that evolution was "clever" to design female menopause (meaning that it has an adaptive origin and function). However, drawing on Gould and Lewontin's spandrel analogy, I pose the question of whether menopause might be incidental (i.e., a figurative spandrel) to constraints and other selective forces.

What if menopause was not considered a trait in and of itself? What if our search for an adaptive function for menopause distracts us from potentially more important evolutionary questions about mismatches between biology and novel environments? What if we created models for the evolution of menopause that incorporated constraints and life course processes that have been selected to (1) slow somatic aging particularly for males who can still reproduce in later life, and (2) optimize female reproductive success earlier in life? What if we viewed menopause as an unavoidable, but incidental event (i.e., a "spandrel") occurring when the rates of reproductive and somatic function and aging (i.e., the "arches") diverged in females due to different evolutionary forces acting on males and females? Would such models and views change our perspective on the function of menopause, and the treatment of it?

While postmenopausal life could be used to invest in grandchildren, just like spandrels are used for beautiful art, evolutionarily when considering both males and females, it makes more sense for older males to invest in mating (producing children that share half their genes) than for grandmothers to invest in grandchildren (that share at most a quarter of their genes). There would be a spillover effect with genes for longevity (or slowed aging) in males with higher reproductive success being passed down to their daughters as well. Older females who choose to invest in grandchildren may derive fitness benefits, but such potential fitness benefits are unlikely to have driven the extension of lifespan or affected timing of female reproductive functioning since that is phylogenetically decoupled from somatic aging and developmentally constrained.

If, like a spandrel, menopause is not an adaptive "design" but an unintentional consequence of selection on other traits such as the number of follicles and rates of atresia to maximize earlier female reproduction plus the extension of lifespan in males, what does that mean for symptoms and health problems? If we don't fully comprehend evolutionary tradeoffs involved in the phenomenon of menopause, do we risk unintended consequences of well-intentioned but misinformed interventions? For example, while birth may not appear "designed" well (in terms of minimal pain and risk to mothers), human babies have evolved to "expect" a vaginal birth and all the mess that goes along with it, especially the compression in the birth canal and the microbes from the vaginal-anal regions they are initially colonized with. Medical interventions like C-sections might at first appear to bypass the risks associated with vaginal births, but they have unintended long-term consequences on babies (e.g., increased asthma and allergies due to different initial microbes) and on mothers (e.g., increased risks for subsequent pregnancies).[2]

In the case of menopause, what if we intervene to extend women's ability to become pregnant by slowing rates of follicular atresia starting at birth to extend women's childbearing years, particularly as women in many populations want to have children later in life? (Perhaps such alternative phenotypes have existed in the past, but had lower reproductive success.) What if we design hormonal contraception to prevent pregnancy without thinking about the body's "expected" hormonal exposure throughout the life course and potential long-term health effects that may only manifest during or after menopause? Interventions for menopause (e.g., Menopausal Hormone Therapy, the newer term for Hormone Replacement Therapy and related hormonal treatments given at menopause) that depart from the body's evolutionary expectations have already demonstrated unintended consequences like those observed in the Women's Health Initiative (WHI) study (see Part I). Treating complex systems with relatively simple interventions, without taking an evolutionary life course perspective, might have unintended consequences for aging and quality of life.

In this chapter, we first explore some evolutionary considerations related to menopause, briefly review several adaptive evolutionary hypotheses, and evaluate comparative evidence for menopause. We then explore whether menopause might be better thought of as a figurative spandrel, incidentally formed by two proposed figurative "arches" that reveal the spandrel of menopause: (1) an extended arch of somatic function and lifespan, linked to reproductive lifespan in males, since they can directly increase their fitness by reproducing at later ages, and (2) an arch of female reproductive function under phylogenetic and developmental constraints to maximize earlier reproduction. Finally, we examine evolutionary mismatches in reproductive ecology and history and how those may impact the experience of menopause and beyond.

10.1 Evolutionary Considerations

The big evolutionary puzzle is why can human females live past menopause, and spend up to half of their adult lives not reproducing if reproductive success is so important? Most human females reproduce between the ages of 15 and 45 (a span of about 30 years), but in contemporary industrialized populations, average female life expectancy at birth is about 80. This means that many women may have post reproductive lifespans as long or even longer than their reproductive lifespans. Related evolutionary questions exist. Is menopause unique to humans? Why did menopause evolve, and from what prior trait? At the origin, were there adaptive benefits conferring a fitness advantage? Or were there other selective pressures, combined with phylogenetic or developmental constraints, that "revealed" menopause? Why would it ever make evolutionary sense to continue living but stop reproducing? Could current experiences of menopause reflect an evolutionary mismatch with contemporary lifestyles?

For selection to act on a trait such as menopause or post-reproductive lifespan, the following criteria must be met:

- population variation in the trait (e.g., some women living past menopause, others not)
- genetic or heritable component upon which selection can act (e.g., there must be genes influencing the rate of reproductive and/or somatic aging)
- competition for survival and reproduction (e.g., for food and similar resources and/or mates and reproductive opportunities)

- differential reproductive success (e.g., between women who live past menopause and those who do not)

Evolutionary biologists have predicted a sharp die-off of individuals near the time of reproductive cessation.[3] But this "wall of death" does not occur in humans. Is that because there are adaptive benefits to staying alive after that period (validating Live-Long Hypotheses[4] – including the Grandmother Hypothesis), or is it because there is minimal cost to menopause and thus selection has little "power" to cull this post-reproductive population of females? Did women who stopped reproducing earlier (Stop-Early Hypotheses[4]) have greater fitness than those who reproduced into their 50s or 60s? Were there ever individuals who could reproduce into their 50s, 60s, 70s? Since we do not have data on variation in reproductive aging existing at the origin of menopause, we can only speculate.

The implied alternative to becoming postmenopausal (i.e., being able to continue reproducing past 50) appears constrained by (1) phylogenetic history, involving separation between reproductive and somatic aging that occurred in ancestors of mammals and (2) development, with ovarian follicle numbers that peak in utero and decline dramatically before reproductive maturity, suggesting some evolutionary optimization to maximize fitness in women's 20s and 30s.

If evolutionary constraints limited our female ancestors' ability to reproduce later in life, then we would need to explain why lifespans got extended, and why women don't die suddenly around menopause. Since selection weakens with age, particularly post-reproduction, it may be hard to select against postmenopausal women.

10.2 Adaptive Evolutionary Hypotheses

Many adaptive hypotheses have been proposed for the evolution of menopause,[5] and several are described below. By definition, postmenopausal women can no longer reproduce, so evolutionary explanations for menopause often consider reproductive success resulting not only directly from one's own offspring but also indirectly from behavior that supports related individuals who share genes (inclusive fitness), particularly investment by grandmothers in their grandchildren.

10.2.1 Grandmother Hypothesis

One of the most widely discussed and tested explanations for a long post-reproductive lifespan is the Grandmother Hypothesis,[6] which posits that women who ceased reproducing and invested in kin had greater reproductive success than women who continued increasingly risky attempts at reproduction. This hypothesis suggests that older women who stopped reproducing earlier, and/or those who outlived their ability to reproduce, gained an evolutionary advantage when they devoted their remaining energy toward increasing fitness of their own children and grandchildren.

Grandmothering has been hypothesized to explain unique human life history traits (e.g., later maturity, lower ratio of weaning weight to adult weight, and higher fertility) and more altruistic behavior compared to other primates.[7] Human life history is characterized by a prolonged juvenile phase during which children are dependent on mothers but also others,[8] including grandmothers, who may assist with knowledge, household tasks, childcare, nutrition, and survival, thereby increasing offspring reproductive success by caring for their grandchildren.[9,10] In Japan, many grandmothers invest in their grandchildren.

Fieldnotes

Evidence of the work grandmothers may do for grandchildren abounds. As part of my study. I took various body measurements, including arm circumference. Two women were outliers in my sample, having extremely large biceps. One was a baker, and spent hours a day kneading bread. The other told me she picked up her grandkids all day long. This comparison provides a measure of the amount of work some grandmothers might provide for their grandchildren, particularly if they are healthy enough to do so.

Yet, the Grandmother Hypothesis seems less than persuasive because it conflates current adaptive benefit (which may appear considerable) with origin (which is less clear). If an evolutionary advantage was gained with the origin of menopause, compared to whom? The implied ancestral condition is either women who could reproduce longer or were somehow "programmed" to die around menopause. We have no evidence for the former, and the idea of programmed aging has been widely criticized (see Chapter 9). Imagining a plausible origin scenario where inclusive fitness of investing in grandchildren would be the primary driver of menopause in ancestral humans is difficult given weaker selection later in life, and males' ability to reproduce later in life, which could drive longevity in their female offspring as well.

Several potential challenges to the Grandmother Hypothesis exist. First, while tradeoffs between investing in reproduction versus somatic maintenance and repair are assumed likely based on Life History Theory, recent studies suggest that, at least in contemporary Western, Educated, Industrialized, Rich, Democratic (WEIRD) populations, reproduction at later ages (as proxy for reproductive health) might actually correlate with longer lifespan.[11] While not necessarily relevant to the origin of menopause, it suggests that genetic variation exists for possibly extending reproductive lifespan. However, it may also be that in WEIRD populations, other factors that influence reproductive health and lifespan also influence longevity, contributing to such a correlation (Figure 2.2).

Second, paternity uncertainty poses a challenge to the Grandmother Hypothesis, if grandmothers invest in their sons' children. Unlike females who can be certain that children they carry are genetically related (i.e., maternal certainty), paternity uncertainty results from males' inability to know for certain that their sperm fertilized their partner's egg (unless they control access to females). When paternity certainty is low, males (and their mothers) who take care of their partner's offspring might be investing in competitors' offspring who do not share their genes. Evolutionarily, a tendency to invest heavily in offspring who may not be theirs would be selected against because their genes would have decreased representation in subsequent generations.

Third, and relatedly, in societies in which males stay resident and females migrate out (patrilocal pattern), which comprise most contemporary human societies as well as those of our nearest primate relatives and thus likely our ancestors, grandmothers who invest in grandchildren (usually of their sons) can never be certain they are ensuring the survival of their own genes. In such societies, because of out-migration of females, most grandmothers would be paternal grandmothers.

Fourth, we should not assume that all grandmothers provide net benefit to their children and grandchildren. Best-case scenarios, with evidence of grandmaternal investment in grandchildren, do exist:

Fieldnotes

Following tradition, one participant and her husband, a first-born son, lived with his mother. While many women complained of tension with their mothers-in-law, this woman told me how fortunate she felt that she was able to continue working while her mother-in-law cared for the home and her children, even when she worked night shifts. Later, despite not having sons of her own, she cared for her daughters and their children whenever they came to stay (which they did often and for extended periods of time). She also supported many other women with young children through her extensive volunteer activities. This may be an example of how the tendency to care for one's own kin may "spill over" into caring for children of non-relatives, or altruism that is seen among human societies and poses an evolutionary puzzle.

But many women's realities differed from this ideal case. In Japan and other cultures, the stereotype of difficult mothers-in-law who might impose burdens on their families, is widespread. Several participants talked of difficult mothers-in-law, and one woman even told me she went to great lengths to avoid marrying a first-born son to avoid the familial obligations, including living with her in-laws, only to have her husband become a *de facto* first-born son after their marriage. Depending on the physical and mental health of a grandmother, the net transfer of resources could go in the opposite direction from that predicted by the Grandmother Hypothesis, and possibly even compromise reproductive success. Burdens of dealing with, and possibly caring for, some grandmothers might offset their contributions to care of grandchildren in some households, at least later in life. It is important to note that psychological distress does not necessarily translate to negative impact on reproductive success, but it challenges the assumption that grandmothers' presence is always beneficial.

10.2.2 Mother Hypothesis

While investing in grandchildren (who on average share a quarter of a grandparent's genes) would contribute to lifetime reproductive success, investing in children (who on average share half of a parent's genes) would be more advantageous. Given the extended period of human infant and child dependency on parents (particularly mothers), adaptive benefits are expected to accrue to women who survive until their last offspring reach maturity, regardless of their involvement with grandchildren. In contrast to most mammals that become relatively independent when they are weaned, human children remain dependent on mothers (or others) well beyond weaning for protection from predators, food procurement, and acquiring cultural skills and knowledge essential to survival and reproductive success. Thus, living beyond the last child's birth for at least 15–20 years may not require investment in grandchildren to make evolutionary sense, because it is adaptive to invest in one's own offspring until they reach reproductive maturity (otherwise they don't "count" in the reproductive success calculation).

The Mother Hypothesis posits that menopause occurs to ensure prolonged investment in existing offspring by decreasing risk of maternal mortality from subsequent pregnancies. This "extended mothering" (relative to other primates) may be a hallmark of humans. Women in natural fertility populations, and probably throughout our evolutionary history, have children until menopause, as evidenced by observations of women having their last menstrual period prior to their last pregnancy, and moving from pregnancy to lactation to menopause without any intervening periods.[12] Thus, the Mother Hypothesis would explain a post-reproductive lifespan of 15–20 years, with last birth around 45, menopause around 50, and living until around 65, which aligns with the modal age of death of 70 among Tsimane hunter-gatherers.[13]

Mothers have the greatest positive effect on child survival, with fathers and paternal grandmothers showing mixed results (as might be expected given paternity uncertainty and costs of reproduction).[10] Interestingly, data from seven primate species suggests that in some of our nearest relatives, young offspring are more likely to die even while their mother is still alive if she is facing impending death in the next few years,[14] presumably due to her declining health. If healthy mothers are important to their infant's survival for 15–20 years, the phenomenon of women living at least 15–20 years after menopause could be explained by the Mother Hypothesis.

10.2.3 Father Hypothesis

What if evolutionary hypotheses for menopause based solely on female life histories are missing something? Selection acts on all individuals of a species via their offspring, which on average include approximately equal numbers of males and females, suggesting that a consideration of males might be critical. Genes associated with longevity have not been found on the Y chromosome, suggesting they are likely shared by males and females.

Mutations that contribute to lengthening lifespan could increase in a population if men were able to increase their reproductive success by having additional children, without requiring any fitness benefits from grandmothering. Increased survival of men would have a "spillover effect" on women's lifespan because men pass their genes to children of both sexes.[15] Thus, evolutionary hypotheses about female menopause, or the lifespan extensions that reveal it, might benefit from including life histories of males and mating patterns.

Could selection on older males be driving female post-reproductive lifespans? Males often exhibit more variation in reproductive success than females. While male health declines with age, the ability to produce sperm continues throughout life (although sperm "delivery" can be affected by health- and age-related decline) and thus selection may act more strongly on males than females in later life. Male fecundity can be significant after 50, particularly for males with access to multiple female partners. Females can usually only have one child a year, whereas there is no real limit to how many children men can father in a year, as that is determined primarily by their access to females. Selection on male reproduction predicts a gradual rise in mortality risk after age 50, without requiring any "grandmother effects," and this may be the primary selective force breaking the "Wall of Death." Importantly, while classic evolutionary theories for menopause focus on a single (female) sex model, two-sex models that include males (with later life fertility) and allow for mating between older males and younger females do not require female-inclusive fitness (grandmothering) for a post-reproductive lifespan to originate.[16]

The Father (or Patriarch[17]) Hypothesis argues that menopause is a by-product of selection for longevity in males, explained by variable male reproductive success, and males being capable both of maintaining high status (which may increase access to multiple mates) and reproducing into their later years. Evidence comes from contemporary cross-cultural mating patterns that may provide clues to patterns prevalent in ancestral populations including:[15]

- "Mildly" polygynous (males mate with more than one female), based on comparative testes size and sexual dimorphism
- Age gaps at marriage of 5–15 years in traditional societies (and even contemporary societies in subsequent marriages with serial monogamy)
- Gerontocracy (groups governed by older people), with older men monopolizing access to reproductive females.

Male-centered hypotheses introduce sexual selection into the discussion of the evolution of menopause, but might remove potential female agency by focusing solely on male mate choice. In most species, sexual selection involves female choice, with females acting as the primary selective force influencing which genes get into the next generation. In humans, sexual selection may occur when younger women choose to mate with resource-rich older men rather than resource-poor younger men. While cultural constraints on female choice in many historical patriarchal societies may have been widespread, situations in early evolutionary history may have been different, again reminding us of the importance of not extrapolating from recent history to conditions likely present at the evolutionary origin of a trait like menopause.

10.3 Comparative Evidence for Menopause

From a comparative perspective, an extended human female post-reproductive lifespan may be one of the most unusual life history traits in the animal kingdom. But is it *qualitatively* different (with no other animals exhibiting such a life history trait), or *quantitatively* different (distinguished by its relative length but not its existence)? Answering this question may help us identify how far back in our evolutionary history we must go to find the origin(s) of this trait.

Few animals exhibit a prolonged post-reproductive lifespan except for humans, chimpanzees (mostly in captivity), several whales, and some social insects.[18] Thus, humans appear relatively unique not only in terms of having a prolonged post-reproductive lifespan (senescent stage), but also because of their relatively long childhood life stage (compared to other primates).

Comparative analyses of menopause in other species, in which it may have arisen independently (i.e., not due to shared evolutionary history), may also provide insights into environmental contexts and selective pressures. Menopause appears to have evolved several times in multiple species of toothed whales by lifespan extension without increased reproductive lifespan.[4] This would be expected if reproductive lifespan was constrained by developmental physiology and decoupling of somatic and reproductive aging. Offspring are dependent on mothers for many years (at least until they are weaned), such that "extended mothering" in whales and humans may explain some of the post-reproductive lifespan.

Multiple paths could also lead to apparently similar phenomenon of menopause cross-species. For example, perhaps (grand)maternal investment is key in whales, but male ability to father multiple children later in life (particularly as cultural resources and status play a role in access to mates among humans) may be a driving force in humans.

While humans are relatively long-lived, there are mammals with longer lifespans such as the arctic bowhead whale (estimated to live up to 200 years). Other long-lived mammals continue to reproduce until they die, and oocyte stocks show variation among mammals,[18] suggesting potential variation in the mammalian lineage for selection to act on to delay menopause. Many animals exhibit post-reproductive lifespans in captivity, suggesting that variation exists and may be expressed given certain environmental conditions (e.g., reduced disease and predation in captivity). Chimpanzees have been known for some time to go through menopause in captivity and recently it has been shown that this may also happen in the wild,[19] demonstrating the capacity for menopause in our closest relatives given certain environmental conditions. What does it mean if other animals can have menopause in certain environmental contexts?

Importantly, while the apparent "trait"of menopause or post-reproductive lifespan appears relatively unique to humans, and a few other animals, the decoupling of reproductive and somatic senescence predates humans and even mammals and birds. Thus, this decoupling is not a human trait, but something primitive to mammals and birds, and therefore phylogenetically constrained.

10.4 Menopause as Spandrel

While some of the adaptive hypotheses are appealing for their simplicity, their neglect of other evolutionary mechanisms (e.g., constraints and selection on both males and females) is concerning. By assuming adaptive benefits were key to the origin of menopause and a post-reproductive lifespan, we risk falling prey to the "adaptationist program" cautioned against by Gould and Lewontin.[1] We do this at the expense of understanding the complex phylogenetic, developmental, and physiological constraints that have produced a female reproductive system that has contributed to humans becoming a reproductively successful species, as well as potentially more powerful selective forces acting on men. It may seem counterintuitive to include males in evolutionary hypotheses about female menopause, but reproductive success requires both females and males, and results in both. We must take a multigenerational approach to evaluating reproductive success, since that is ultimately what reproductive success is about – contributing to future generations.

Consider the analogy of the architectural spandrel. How might this provide insight into the evolution of menopause? Could the figurative spandrel of menopause be produced, not from selection for a "trait" of menopause, but rather from selection to (1) slow somatic aging particularly for males who can still reproduce in later life, and (2) optimize female reproductive success earlier in life. These are related respectively to (1) the father hypothesis and (2) the mother hypothesis discussed above. Could menopause be more productively viewed as an inevitable event (i.e., a "spandrel") that occurred when the rates of reproductive and somatic function and aging (i.e., the "arches") diverged in females due to different evolutionary forces acting on males for later reproduction, and constraints acting on females to maximize earlier reproductive success (Figure 10.2)?

This architectural model of menopause (Figure 10.2) is of course figurative, as is the idea of the two arches of reproductive function, but the analogy vividly demonstrates how menopause may be an incidental feature in human females, the result of the divergence of both arches rather than a key aspect of the design structure itself.

As depicted in Figure 10.2(a) by overlapping curves of reproduction and somatic functioning, a plausible ancestral trait for human males and females is that reproductive and somatic aging likely followed similar patterns of functioning over the life course. In males these curves (or arches) are linked and overlap since males can reproduce over their entire lifetime. But in females, these curves of reproductive and somatic aging are decoupled (and have been since long ago in our evolutionary history) due to phylogenetic and developmental constraints.

Since males can continue to produce viable sperm throughout adulthood and thus reproduce later in life, selection could favor males who live longer and leave more children. Thus, selection could have acted to slow somatic aging and extend human lifespan in males, and their reproductive and somatic function/aging curves (which are linked) would be extended, forming what appears to be a single arch (Figure 10.2[b]).

On average half of their daughters would have inherited the gene(s) that promoted slower somatic aging. But those daughters would have experienced only slowed somatic aging, since rates of reproductive aging would have been phylogenetically and developmentally constrained, thus leading to the two arches (of reproductive and somatic function/aging) diverging and revealing the "spandrel" of menopause (Figure 10.2[c]).

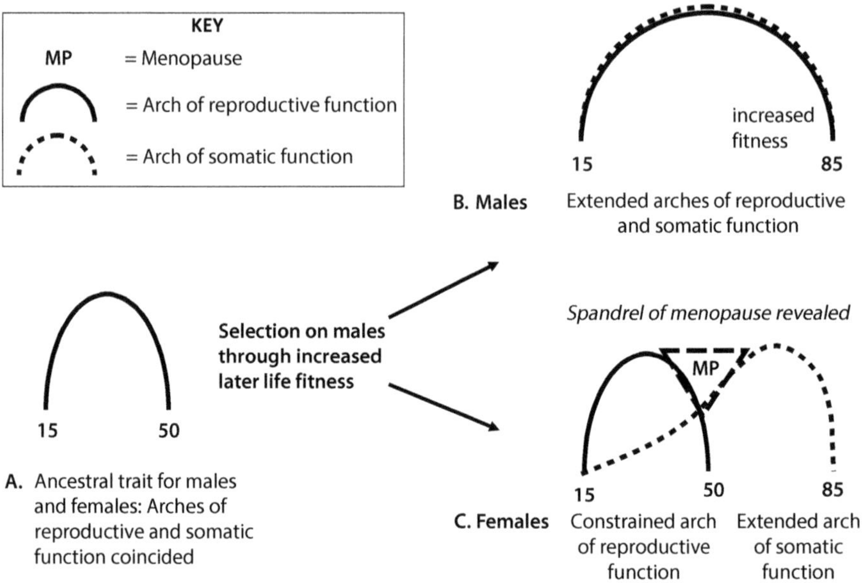

FIGURE 10.2 Schematic outlining the "*Menopause as figurative spandrel*" hypothesis. (a) Starting from a hypothesized ancestral trait, the figurative spandrel of menopause is revealed by (b) selection for extended lifespan and later life reproductive and somatic function in males, and (c) constraints on reproductive function in females optimizing early life reproduction. This leads to divergence of the two figurative arches of reproductive and somatic function/aging thereby revealing the "spandrel" of menopause. (Numbers under arches refer to age in years)

This could be considered a metaphorical spandrel, revealed by the two metaphorical arches of (1) extended somatic function (or aging) selected for in males, since somatic and reproductive function in males overlap, and (2) constrained reproductive function in females. Some females might use the "spandrel" of menopause or post-reproductive lifespan to increase reproductive success by investing in grandchildren, but that would not be driving the origin of menopause. This post-reproductive life stage does not have to be used for reproductive success, aligning with mixed results of studies on adaptive benefits of grandmaternal investment in different populations, particularly given paternity uncertainty and residence patterns that may make it hard to calculate.

10.4.1 Evidence for Arch 1: Extended Somatic Function Leading to Increased Male Fitness

If at some point in our evolutionary history, a mutation arose in a gene that regulated the timing of development or somatic repair and maintenance such that somatic aging slowed, a male with such an allele might have been able to have additional children during that extended lifespan (since somatic and reproductive aging coincide). Both his male and female offspring would have a 50% chance of receiving this mutated gene, since genes related to aging mostly appear to be located on non-sex chromosomes.

Male reproductive success tends to be more variable than female reproductive success and depends primarily on access to mates. And males can have multiple mates, as many human societies appear polygynous (with one male mating with multiple females). While females are limited to a single offspring per year (unless they have multiple children in a single birth), in theory, a male could father tens or even hundreds of children during that time frame, limited primarily by access to mates. The maximum recorded individual lifetime fertility rate is 69 for a female from births involving multiple children (e.g., twins or triplets), many of whom do not usually survive,[20] compared to hundreds or thousands for a man named Genghis Khan who had access to multiple female mates and appears to have contributed disproportionately to subsequent generations.[21]

In less extreme cases than Genghis Khan, females may choose to mate with older men who may be more likely to have resources. Such female choice may be an example of sexual selection. In our evolutionary history, resources might have included social capital and/or hunting skills to procure food. Mating patterns cross-culturally often show age differences with males being older than females, and the age difference is often larger for second and subsequent wives. Thus, there is historical and contemporary evidence of males mating with multiple and with younger females, and thus receiving fitness benefits at older ages. A model involving both males and females demonstrates that selection might favor survival for as long as men can reproduce, supported by data that males over 50 increase their fitness by mating with younger females.[15]

For females, the immediate costs of reproduction involving pregnancy and lactation are high; for males, reproductive costs are very low because sperm are "cheap." Thus, the evolutionary cost-benefit analysis of investing in reproduction versus maintenance of somatic tissue differs for males and females throughout the life course. Variation in male reproductive success, variation in mating patterns, and cost-benefit ratios of investment in reproduction lay the groundwork for fitness benefits to easily accrue to males who have mutations allowing them to live longer. And if there were no appreciable fitness costs to extending this later stage, it would not have been selected *against*, in male or female offspring. As lifespan increased, because reproductive aging is more tightly linked with somatic aging in males, the arches of reproductive and somatic aging would both have been extended for males, but not for females.

10.4.2 Evidence for Arch 2: Constrained Reproductive Function in Females Optimizing Fitness

From an evolutionary history (phylogenetic) perspective, the potential for female somatic lifespan to extend beyond reproductive lifespan is not novel in humans. Decoupling of female rates of somatic and reproductive aging (meaning that rates of aging differ for reproductive and somatic organs and tissues in females) occurred much earlier in evolutionary history, when the ancestors of mammals and birds diverged from other animals in the tree of life. This decoupling would allow somatic aging to be extended while reproductive aging was constrained by phylogenetic history.

Fieldnotes

While evolution may have separated reproductive from somatic aging in female mammals and birds, humans have also figured out how to remove the process of reproduction from the body with various reproductive technologies like sperm donation and in vitro fertilization. One woman asked me if I wanted children. At the time I wasn't sure and said so. She then proceeded to advise me that if I wanted a child, "seishi wo katta hō ga ii." *I didn't know the word* seishi *and looked it up in the dictionary. I was shocked to realize that she had just told me "it would be better to just buy sperm!" When I followed up, she explained that she thought most men were children (at least her husband anyway), and thus unless I wanted a grown man who would act just like a child, or need the care of a child, I should just buy sperm. I asked if that were possible in Japan, and she said no, but that she had heard it was possible in the United States!*

Sperm is relatively easy to come by. In contrast, eggs, and the females who nurture fertilized eggs through months of pregnancy, are considered the rate-limiting step of reproduction. Female reproductive development and aging are tightly constrained and have been since long ago in our evolutionary history. In females, the statistical probability of getting pregnant and having a live birth (fecundity) exhibits an inverted U-shaped pattern from puberty to menopause. This maximizes reproductive success since it is better to reproduce earlier, once maturity is reached. In the early years, the ability to get pregnant is low while girls are completing pelvic growth and accumulating fat stores to support pregnancy and breastfeeding. It increases gradually until a woman's 20s and 30s. In later years, decreased follicle number and changes in the hypothalamic-pituitary-gonadal axis lead to decreased fecundity. If lifespan is long enough, fecundity decreases to zero by menopause.

This reproductive architecture is likely highly constrained by evolution and phylogenetic history, as well as underlying developmental and physiological constraints of the hypothalamic-pituitary-gonadal axis (see Part I).

Cross-cultural patterns of female hormone levels are similar over the reproductive life course despite lower absolute levels in non-WEIRD populations.[22] This suggests evolutionary optimization of the inverted U-shape pattern of hormone function over the reproductive life course. The variation in absolute hormone levels appears influenced by biocultural reproductive ecology factors. The robust cross-cultural patterns for reproductive functioning likely result from optimizing reproduction between the ages of 20 and 40, and influence why menopause occurs when it does (i.e., when the number of follicles and associated hormones fall below a certain level at the tail end of the inverted U-shaped curve). The higher absolute levels of hormones (and potentially more dramatic hormonal changes during the menopausal transition) in women in industrialized populations might provide a reason for why they experience more significant menopausal symptoms (e.g., perhaps due to departures in environment, lifestyle, and physiology from the evolutionary norm for our species, which lead to higher exposure to hormones).

Hormones regulate the timing of development as well as adult reproduction and are involved in mechanisms of biological aging.[23] Genes and environment influence the initial stock of follicles containing eggs and their rates of depletion by follicular atresia in order to optimize reproductive hormone function in women during their 20s and 30s. The processes that contribute to optimization of reproduction early in the life course, likely also influence depletion of those eggs by the early 50s, thereby resulting in menopause. This may be an example of antagonistic pleiotropy, whereby genes for maximum oocyte stock and rates of follicular atresia have positive effects on fitness early in life, but then lead to potential "negative effects" associated with menopause later in life.

Since human menopause occurs when follicular number decreases below a certain threshold (~1,000 follicles), the original stockpile of follicles (which reaches its maximum at about five months of the woman's own gestation) and the rate of follicular atresia (degeneration of ovarian follicles, which occurs throughout life) may constitute developmental constraints.[24] These follicles, and the rate of follicular atresia, not only influence the timing of menopause[25] but also the physiological functioning and feedback loops of the brain-ovarian system (hypothalamic-pituitary-gonadal axis), which are critical in influencing reproduction. Thus, they are likely to have been subjected to significant selection to maximize reproductive success.

When considering evolutionary hypotheses, it is important to consider alternative phenotypes that may have arisen such as women who could continue reproducing past age 50. In theory, mutations could have plausibly arisen to slow rates of atresia, thereby increasing the pool of ovarian follicles at puberty, and delaying the age of menopause. But might that have affected reproductive functioning in negative ways that compromised reproductive success in early years? Starting puberty with too many follicles or extending and flattening the curve of reproductive functioning might have negative effects on lifetime reproductive success. Perhaps such mutations have arisen in our ancestral past. If the rates of atresia and numbers of follicles have been selected for over human (and earlier mammalian) history to optimize earlier reproductive functioning, individuals with such mutations might be unable to outcompete other individuals. While they might be able to reproduce later in life, if this involved tradeoffs with earlier reproduction, this would not be "worth it" in terms of reproductive success. Furthermore, risk of maternal mortality increases with age, and maternal death negatively impacts surviving children, and thus overall reproductive success. This is not true in males, who incur minimal risks to themselves or their existing children by additional matings as they get older. Therefore, as shown in Figure 10.2(c), if the arch of somatic aging was extended, the arch of reproductive functioning could not "follow" for females without potentially affecting early life reproduction and survival of existing children, which would have significant negative effects on reproductive success. Thus, the arches would diverge, thereby creating the space (spandrel) for menopause to occur.

10.4.3 Spandrel Summary

In summary, males invest relatively little in reproduction (insemination at minimum), so they can increase their fitness with relatively little cost. Females invest considerable resources in reproduction (nine months of pregnancy and evolutionarily years of lactation at minimum). While grandmothers can invest time and energy in grandchildren, such effort would only confer fitness benefits if the grandchildren were related, which would be more certain through their daughters, less so through their sons. And others, besides mothers and grandmothers, such as fathers and siblings (who are more related than grandmothers) can care for children.[8] Given the relative costs of reproduction, older males who reproduced later in life would have the potential for considerable fitness advantages (high benefit to cost ratio, potential to father children who share half their genes), and much greater than older females who invested energy into grandchildren (who share only a quarter of their genes, and may or may not even be related if they are through their sons).

Just like spandrels are used for beautiful art, postmenopausal life could be used to invest in grandchildren (Grandmother Hypothesis), and of course mothers are key to child survival and ultimately reproductive success thus requiring them to live for many years post-birth (Mother Hypothesis). Yet, evolutionarily, if we consider both males and females, it makes more sense for older males to invest in mating, with a spillover effect that their genes for longevity (or slowed aging) would be passed down to their daughters as well. Distinguishing between factors influencing the origin of a trait like menopause or post reproductive life span vs. factors influencing contemporary fitness benefits is important, not just theoretically but also in terms of applications.

How do our views on the evolution and function of menopause influence our views and approaches to symptoms and health problems? For example, approaches to extending women's ability to become pregnant later in life by slowing rates of follicular atresia might have unintended consequences. Should we consider how the design of hormonal contraception to prevent pregnancy, without consideration of the body's evolutionarily "expected" hormonal exposure throughout the life course, might impact potential long-term health effects that may only manifest during or after menopause? How might our confusion or conflation of evolutionary origins vs. contemporary fitness benefits obscure mismatches that might result in unintended consequences for aging and quality of life?

10.5 Evolutionary Mismatches

In Part I, we discussed the "normal" menstrual cycle which is expected to occur every 28 days from puberty until menopause, begging the question: normal for whom? We can ask the same question about the experience of menopause. Understanding the environmental factors that influence the entire process of reproductive function may shed light on the possible adaptiveness of its end (i.e., menopause), and have implications for experience and treatment.

Reproductive ecology examines factors influencing reproductive functioning and fertility, often with an implicit evolutionary perspective. To better understand how our current environments and lifestyles influence reproduction (and menopause), and may be mismatched with our evolutionary heritage, we can compare contemporary industrialized populations with hunter-gatherers who may offer a first approximation of the lifestyles and life histories of our ancestors. Contemporary hunter-gatherers are not evolutionary relics from our ancient past but have continued to adapt as have industrialized populations, and their lifestyles have been impacted by global and local changes, ranging from political-economic changes to climate change. However, this approach can provide rough estimates and useful thought experiments.

Eaton et al. compared "natural fertility" (not using medical contraception) hunter-gatherers (as a proxy for our evolutionary ancestors) and Americans on reproductive history factors that are associated with reproductive cancers (Table 10.1).[26] They suggest that changes in our behavior and biology induced by biocultural changes, like delayed marriage and birth,

TABLE 10.1 Typical reproductive experiences and lifestyles of women in evolutionarily "normal" natural fertility populations (using hunter-gatherers as proxies) vs. industrialized populations

	Evolutionarily "Normal" Hunter-Gatherers	*Industrialized Populations*
# lifetime menstrual cycles	163	468
Age at menarche	Late	Early
Age at marriage and first childbirth	Early	Late
Number of children	Many	Few
Breastfeeding duration	Long	Short
Physical activity level	High	Low
Dietary diversity	Great	Limited

Modified from Eaton et al.[26]

birth control (often hormonal), smaller completed family size, and shorter lactation duration, have increased our later life reproductive cancer risk by increasing the number of menstrual cycles and hormonal exposure over the lifespan. Could these same changes impact the experience of menopause and postmenopausal symptoms and disease risks? Could a potential mismatch have implications for how we prevent or treat symptoms associated with menopause?

Figure 10.3 highlights how cultural factors such as age at marriage may have significant consequences on reproductive success.

Fieldnotes

In human reproductive ecology, age at marriage is an important determinant of lifetime fertility in societies where children born out of wedlock are uncommon. Many of my participants were concerned about their kids getting married. Some participants organized an omiai *(arranged marriage meeting) between their children and potential marriage partners, including distant relatives, thus allowing them to draw on their family networks. Some participants even asked me to find someone (preferably foreign males) for their daughters. Many were concerned that their sons and daughters were still living at home, becoming* furītā *("freeter," anglicized; a Japanese term for unemployed or underemployed, non-regular wage earners), with no prospects of marriage or children.*

Variation in age at marriage, contraceptive and breastfeeding practices, and energetics lead to significant differences not only in completed fertility, but also numbers of menstrual cycles, which may have both biological and cultural implications for the significance of the end of menstruation, i.e., menopause.

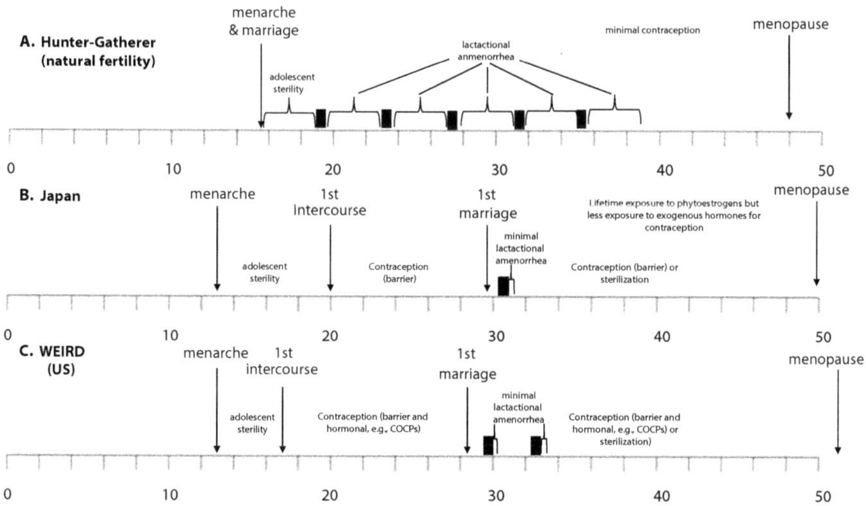

FIGURE 10.3 Population variation in reproductive ecology with estimates of timing of key reproductive events and exposures for: (a) Hunter-gatherers (natural fertility); (b) Japanese; and (c) Western, Educated, Industrialized, Rich, Democratic (WEIRD) populations such as the United States. (Hunter-gatherer data were adapted from Short.[27] Japanese and US data were based on most recent data available online for median [or average when median not available] age at menarche, first sexual intercourse, first marriage, first birth, duration of breastfeeding, and menopause, with most prevalent forms of contraception listed.) Black rectangles indicate pregnancies. COCPs = Combined Oral Contraceptive Pills

Consider a woman living in a typical industrialized population who can expect to have her first menstruation around age 12, and her last menstrual period around age 50. Thus, she would experience menstruation for 38 years. If we assume a "normal" 28-day cycle, that gives 13 cycles/year, and 38 years of 13 cycles/year would give 494 cycles during her reproductive life. Fertility has been decreasing in industrialized populations, in part due to (evolutionarily novel) hormonal birth control. Breastfeeding duration has also decreased due to birth practices that make breastfeeding initiation challenging, lack of social and economic support, inadequate maternity leave, and infant formula marketing and availability. So, if a woman has two pregnancies and breastfeeds for a few months, we can assume that she "loses" ten 28-day cycles with each pregnancy and perhaps three cycles with breastfeeding. So, her lifetime total is 494 – (2 * 13) = 468 cycles. Spread over 38 years, this is an average of 12.3 cycles/year, or roughly "monthly" cycles throughout her entire reproductive lifespan.

In contrast, consider a woman from a natural fertility population, more akin to our evolutionary norm. Data from contemporary hunter-gatherers suggest an average age of menarche at about 16 years, first birth at 19 years, duration of lactational amenorrhea due to breastfeeding of about 2.5 years per child, and a completed family size of 6 children.[27] What does that mean for the number of menstrual cycles over women's lifetimes and in an average year? If the last menstrual period occurs around age 47, that means 31 years of 13 cycles/year for a possible total of 403 cycles. For each of 6 pregnancies with 2.5 years of lactational amenorrhea due to breastfeeding, a woman "loses" 10 cycles with pregnancy, and 30 cycles with breastfeeding, so 40* 6 = 240 cycles "lost" to reproduction. There may be more cycles lost to pregnancies resulting in miscarriage and stillbirth, as six is the completed family size and variable cycles lost to breastfeeding, depending on frequency of nursing and energetics. Thus, the lifetime total would be 403 – 240 = 163 cycles, or maybe less with miscarriages and seasonal nutritional stress or physical activity-induced suppression. Spread over 31 years, this is an average of 5 cycles/year, although many years have no cycles due to pregnancy and lactation.

Interestingly, the medical term "oligomenorrhea" is defined as the "abnormal" condition of "infrequent menstrual periods" of fewer than 6–8 periods/year. Given data on the average number of menstrual periods of hunter-gatherers (average of 5 cycles/year), couldn't 6–8 periods/year be viewed as evolutionarily excessive? This provides an example of medicalization, the defining and treating of "normal" conditions as medical issues (see Part IV). For women in such natural fertility populations, with an average of six kids, and breastfeeding duration of three years or more, it is likely that some might lactate into menopause and thus not even notice it.

In WEIRD populations, the end of menstruation after experiencing roughly monthly cycles for almost 40 years might be more noticeable and more culturally relevant, compared to women who only experience a third of those. Biologically, a woman's body might be more "used to" such cycles and the sudden change or lack of them, might also have greater biological impacts. And if having three times as many menstrual cycles increases reproductive cancer risks because of elevated hormonal exposures,[26] it seems plausible that it might also impact the experience of menopause and other post-reproductive health conditions.

The idea that it is "normal" to have monthly menstrual cycles from menarche to menopause not only appears evolutionarily weird, but also has influenced hormonal medical interventions not only for menopause, but also for younger women during the reproductive years. The design of early birth control pills[28] was influenced by what the creators thought was "normal:" 28-day cycles with a week of bleeding each month. So, the birth control pill regimen consisted of 21 daily hormonal pills, followed by seven days of a placebo or no pills, during which women experience withdrawal bleeding.

For decades, many women were put on birth control pills to regulate (i.e., "make monthly") their irregular cycles. Would that have happened if more people understood the evolutionarily normal pattern of relatively infrequent monthly cycles? Newer hormonal birth control methods often prevent menstrual cycles from occurring for six months or longer, more akin to time durations associated with pregnancy and subsequent lactation. These may appear more evolutionarily normal from the perspective of menstruation, although the underlying reasons (i.e., pregnancy or lactation versus exogenous hormones) and associated hormonal exposures differ significantly, with long-term effects we do not fully understand.

Fieldnotes

Currently in Japan, there is a cultural ideal of one year being the optimal duration for breastfeeding. In fact, some women take hormones to stop their milk supply after only breastfeeding for one year, despite the human evolutionary breastfeeding norm being three to seven years.[29] *Thus, they take hormones to stop the evolutionary normal process of lactation, an example of a medical intervention that runs counter to the evolved biological system. Interestingly, one participant's mother told me that in previous generations women used to nurse their last child until they went to elementary school (around age 6), demonstrating rapid cultural shifts in maternal-child interactions and expectations that have impacts on the reproductive system and possibly lifelong hormonal exposures for women, not to mention impacts on their children.*

Could such changes in breastfeeding duration, as well as contraception and number of pregnancies, influence not only reproductive cancer risk, but also menopause symptoms and associated health risks? Understanding how our contemporary environment might be mismatched with our evolutionary history can inform cultural attitudes and medical approaches, with implications for how we might view and manage symptoms and health risks associated with menopause.

Maybe the more important evolutionary question is *NOT* why does menopause exist, but why does menopause seem so difficult for so many women in contemporary Western industrialized populations? What if the desire to find an adaptive explanation for menopause is influenced by the fact that most of the researchers live in industrialized populations where menopause seems to have measurable health costs (e.g., osteoporosis, heart disease, etc.)? What if symptoms are mostly an artifact of industrialized environments and contemporary lifestyles, rather than from menopause per se? What if views on the evolutionary function of grandmothers are influenced by cultural beliefs that women's roles are primarily reproductive, or serve to reinforce those roles through medicalization? These are some of the questions we will examine in Part IV. And as we move to considering Ultimate Cultural perspectives in Part IV, it is interesting to reflect on what aspects of culture might influence our evolutionary hypotheses focusing preferentially on fitness benefits of grandmothers instead of developmental and phylogenetic constraints, all of which are part of evolution. Is it difficult to escape from the cultural trope of women as "reproducers" even after their biological ability to reproduce has ended?

References

1. Gould SJ, Lewontin RC. 1979. The spandrels of San Marco and the Panglossian paradigm: A critique of the adaptationist programme. *Proc R Soc Lond B.* 205:581–598. https://doi.org/10.1098/rspb.1979.0086
2. Ríos-Covian D, Langella P, Martín R. 2021. From short- to long-term effects of C-section delivery on microbiome establishment and host health. *Microorganisms.* 9(10):2122. https://doi.org/10.3390/microorganisms9102122
3. Hamilton WD. 1966. The moulding of senescence by natural selection. *J Theoret Biol.* 12(1):12–45. https://doi.org/10.1016/0022-5193(66)90184-6
4. Ellis S, Franks DW et al. 2024. The evolution of menopause in toothed whales. *Nature.* 627(8004):579–585. https://doi.org/10.1038/s41586-024-07159-9
5. Perls TT, Fretts RC. 2001. The evolution of menopause and human life span. *Ann Hum Biol.* 28(3):237–245. https://doi.org/10.1080/030144601300119052
6. Hawkes K, Coxworth JE. 2013. Grandmothers and the evolution of human longevity: A review of findings and future directions. *Evol Anthropol: Issues News Rev.* 22(6):294–302. https://doi.org/10.1002/evan.21382
7. Hawkes K, O'Connell JF et al. 1998. Grandmothering, menopause, and the evolution of human life histories. *Proc Natl Acad Sci.* 95(3):1336–1339. https://doi.org/10.1073/pnas.95.3.1336
8. Hrdy, S.B. (2009) Mothers and *others: the evolutionary origins of mutual understanding.* Cambridge, MA: Belknap Press of Harvard University Press, 422.
9. Hawkes K, O'Connell JF, Blurton Jones NG. 1997. Hadza women's time allocation, offspring provisioning, and the evolution of long postmenopausal life spans. *Curr Anthropol.* 38(4):551–577. 10.1086/204646
10. Sear R, Mace R. 2008. Who keeps children alive? A review of the effects of kin on child survival. *Evol Human Behav.* 29(1):1–18. https://doi.org/10.1016/j.evolhumbehav.2007.10.001
11. de Kat AC, Roelofs F et al. 2024. Late reproduction is associated with extended female survival but not with familial longevity. *Reprod BioMed Online.* 49(3):104073. https://doi.org/10.1016/j.rbmo.2024.104073
12. Beyene, Y. 1986. Cultural significance and physiological manifestations of menopause: A biocultural analysis. *Culture, Medicine and Psychiatry.* 10:47–71.
13. Gurven M, Stieglitz J et al. 2017. The Tsimane Health and Life History Project: Integrating anthropology and biomedicine. *Evol Anthropol.* 26(2):54–73. https://doi.org/10.1002/evan.21515

14. Zipple MN, Altmann J et al. 2021. Maternal death and offspring fitness in multiple wild primates. *Proc Natl Acad Sci.* 118(1):e2015317118. https://doi.org/10.1073/pnas.2015317118
15. Tuljapurkar SD, Puleston CO, Gurven MD. 2007. Why men matter: Mating patterns drive evolution of human lifespan. *PLoS One.* 2(8):e785. https://doi.org/10.1371/journal.pone.0000785
16. Morton RA, Stone JR, Singh RS. 2013. Mate choice and the origin of menopause. *PLoS Comput Biol.* 9(6):e1003092. https://doi.org/10.1371/journal.pcbi.1003092
17. Marlowe F. 2000. The patriarch hypothesis: An alternative explanation of menopause. *Hum Nat.* 11(1):27–42. https://doi.org/10.1007/s12110-000-1001-7
18. Croft DP, Brent LJN et al. 2015. The evolution of prolonged life after reproduction. *Trends Ecol Evol.* 30(7):407–416. https://doi.org/10.1016/j.tree.2015.04.011
19. Wood BM, Negrey JD et al. 2023. Demographic and hormonal evidence for menopause in wild chimpanzees. *Science.* 382(6669):eadd5473. https://doi.org/10.1126/science.add5473
20. Most prolific mother ever. 2025. Available at: https://www.guinnessworldrecords.com/world-records/most-prolific-mother-ever. Accessed 8 Feb, 2025.
21. Zerjal T, Xue Y et al. 2003. The genetic legacy of the Mongols. *Am J Hum Genet.* 72(3):717–721. https://doi.org/10.1086/367774
22. Vitzthum VJ. 2009. The ecology and evolutionary endocrinology of reproduction in the human female. *Am J Phys Anthropol.* 140(Suppl 49):95–136. https://doi.org/10.1002/ajpa.21195
23. Finch C, Rose R. 1995. Hormones and the physiological architecture of life history evolution. *Quart Rev Biol.* 70:1–52. https://doi.org/10.1086/418864
24. Melby MK, Lampl M. 2011. Menopause, A biocultural perspective. *Annu Rev Anthropol.* 40(1):53–70. https://doi.org/10.1146/annurev-anthro-081309-145641
25. Wallace WHB, Kelsey TW. 2010. Human ovarian reserve from conception to the menopause. *PLoS One.* 5(1):e8772. https://doi.org/10.1371/journal.pone.0008772
26. Eaton SB, Pike MC et al. 1994. Women's reproductive cancers in evolutionary context. *Q Rev Biol.* 69(3):353–367. https://doi.org/10.1086/418650
27. Short RV. 1976. The evolution of human reproduction. *Proc R Soc Lond B Biol Sci.* 195(1118):3–24. https://doi.org/10.1098/rspb.1976.0095
28. Gibson M. 2015. The long, strange history of birth control. *Time.* Available at: https://time.com/3692001/birth-control-history-djerassi/.
29. Dettwyler KA. 1995. A time to wean: The hominid blueprint for the natural age of weaning in modern human populations. In Stuart-Macadam, Patricia Dettwyler, Katherine A. (eds): *Breastfeeding: Biocultural Perspectives.* New York: Aldine De Gruyter. https://doi.org/10.4324/9781315081984.

PART IV
Ultimate Culture

11

ULTIMATE CULTURE

Introduction

The cultural organization of the life course may profoundly affect the meanings and experience of aging and menopause. Ultimate cultural factors that may contribute to variation in the experience of and expectations surrounding aging and menopause include: attitudes toward and expectations about roles and responsibilities of men and women particularly as they age; healthcare-seeking behavior; cultural perceptions of aging and menopause; symptom-reporting behavior (see discussion of *gaman* (endurance) in Chapter 6); and lifestyle factors that are socioeconomically constrained.

Ultimate cultural perspectives involve historical and sociocultural factors which contribute to cultural norms and values, examined on larger geographical and longer time scales than proximate cultural explanations, which focus more on individual, subjective, insider/*emic* behaviors and experiences. As discussed in Chapter 6, the *emic*/*etic* distinction often implies Western positionality of the author and readers, which may be associated with various cultural assumptions about aging and menopause. In this chapter, we expand our focus on sociocultural factors influencing the expectations and experience of aging and menopause, contrasting Western, Educated, Industrialized, Rich, and Democratic (WEIRD) populations with Japanese populations. Japan differs in that it is *Eastern*, Educated, Industrialized, Rich, and Democratic, i.e. EEIRD, and has a more collectivist culture compared to the more individualistic cultures characteristic of many Western populations. Thus, in Part IV, we explore how perspectives and insights from Japan may challenge some WEIRD biomedical "universals" about aging and menopause (Figure 11.1).

Behavioral variables such as diet and activity influence aging processes, but focus on these often leads to disproportionate emphasis on individual

DOI: 10.4324/9781003644521-15

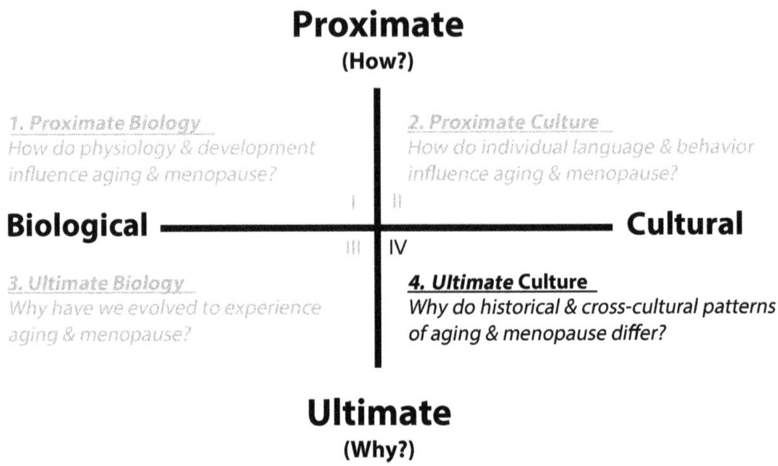

FIGURE 11.1 Ultimate Culture Quadrant IV of the Proximate-Ultimate Biocultural Framework introduced in Chapter 1.

responsibility and sometimes "blaming the victim." Importantly, such behavioral variables are influenced and constrained by ultimate cultural and social/political/economic environments over which people may have little control. In Japan, social determinants of health (SDOH) that likely play important roles in the experience of aging and menopause include: relatively equitable and high standard of living, robust social safety net; universal healthcare; preventive public health screenings; respect for and appreciation of elders; gender roles; and policies and environments that support more proximate healthy individual lifestyle behaviors (e.g., diet and exercise).

11.1 WEIRD vs. Non-WEIRD Societies: Challenging Biomedical "Universals"

Western perspectives are dominant in much of biomedicine (often considered to be the outsider/*etic*, more objective perspective). They reflect the fact that most studies have been conducted in a relatively small proportion of the human population, which may lead to significant bias about the variation that exists in humans. Anthropologists often study non-Western cultures to elucidate insider/*emic* perspectives and communicate them to their mostly Western outsider audiences.

Henrich et al. have argued that it is not just Western, but more specifically WEIRD perspectives that dominate science.[1] The acronym WEIRD conveys that these populations are "weird" in that they depart from cross-cultural norms, as in not reflecting or representing the majority of human experience. People living in such societies ("WEIRDos") are often assumed to be standard and representative of all humans.[1] WEIRDos represented up to 96% of study participants in psychology studies conducted between 2003 and 2007, but only 12% of the world's population, and are not only unrepresentative of humans as a species, but are also outliers on many measures.[1] Furthermore, American college students are oversampled in many studies, comprising over 90% of psychological studies on prejudice, for example.[1]

Compared to other populations, WEIRDos are: individualistic (associated with being overconfident, self-obsessed, and more suicide-prone) and value autonomy; analytical, focusing more on individuals than relationships and context (associated with being less willing to make individual sacrifices for others); and vulnerable to certain optical illusions.[1] Might WEIRDos be vulnerable not only to optical illusions but also other illusions (e.g., involving concepts of success, happiness, and health) with profound implications for quality of life and aging?

Studies of aging conducted among individualistic WEIRDos may thus bias our understanding of the relative importance of individual behaviors and proximate cultural factors. In contrast, might non-WEIRDos help us understand the potential role of ultimate cultural factors? In less individualistic societies, ultimate cultural factors might not only be more apparent, but might also play a larger role. The bias of many studies of health and aging toward WEIRDos implies that we may only have studied a small slice of a much bigger pie of human experience and aging. The rest of the pie might hold important lessons for how we can support healthy aging.

Anthropologists have been studying non-WEIRDos since the beginning of the discipline, recognizing extensive variation among humans and challenging assumptions that Western populations constitute the human norm. Views on sleep provide an insightful example. In WEIRD cultures, sleep is very private, and anthropologists know little about it because we tend to only interact with people when they are awake. However, my experiences in Japan showed me that the boundary between public and private in the context of sleep differs from that of many WEIRDos. Sleeping together is an activity that friends (and even strangers in mountain huts) do in Japan.

Fieldnotes

One time, a participant invited me on a trip with her husband, and another couple. We went to a ryokan *(traditional Japanese inn), where the custom is to arrive in the late afternoon, soak in* onsen *(hot spring baths, usually sex-segregated these days), have a decadent dinner, and then return to the room to sleep – all together. Japan may be one of the few industrialized field sites where an anthropologist can confess that I "slept with two of my participants and their husbands" and not raise any ethical eyebrows. That statement sounds shocking to most Western biomedical researchers, and certainly in the United States where the bedroom is sacred and reserved for sex. Similar statements could be made about bathing naked with participants at public baths.*

Just as this story may challenge some readers' sense of propriety, it may also challenge the individualism inherent in WEIRD societies.

11.2 Japan as Non-WEIRD or (E)EIRD Society

Japan is highly educated, industrialized, rich, and democratic, but not Western, thereby making it EIRD, or EEIRD, for Eastern, Educated, Industrialized, Rich, and *in Japan* democratic. Could there be something about Western vs. Eastern (or at least Japanese) culture that contributes to differences in the experience of aging and menopause?

Could collectivist perspectives of EEIRD populations like Japan help us to better understand social influences on health? Social stress can lead to psychological symptoms,[2] which in WEIRD cultures tend to be dismissed as being "in your head" and somehow less real than symptoms that are "in your body," reflecting the mind-body dualism common in the West which may contribute to limited understanding of psychophysiology. In contrast, in many non-WEIRD cultures, symptoms resulting from social tensions might be considered paramount. In collectivist cultures such as Japan, individual proximate symptoms of aging and menopause may be more difficult to separate from ultimate societal factors and experiences. Integrating such perspectives may be critical to understanding the variation in psycho(social)physiology of aging cross-culturally. Might the biomedical term "psychophysiology" bias us toward focusing more on individuals and reinforce Western mind-body dualism?

According to Hofstede's model of six cultural dimensions,[3] compared to many Western countries, Japan is considered to be less individualistic and more collectivist. Compared to most Westerners, most Japanese people place higher value on relationships and group identity. For example, during the COVID-19 pandemic, US individualism was evidenced by many US residents being unwilling to wear masks (seen as personal burdens). In contrast, in pre-pandemic Japan, people often wore masks in public when they had common colds, so as not to spread them to others. After COVID, many people continued to mask even outdoors, likely influenced by collectivist cultural norms.

Fieldnotes

When my son had field day at his elementary school, I was struck by the lack of individual awards; only team awards were given. It was not that Japanese children were not competitive. Rather, they competed at the level of groups – in stark contrast to my experiences of field days in the United States. In Japan, the teams seemed to be chosen to be relatively balanced in terms of athletic abilities, including children with visible disabilities. There was no indication of resentment by other team members of those less skilled at games – they were all working together to improve their team scores.

How might this collectivist philosophy, exhibited at field day, influence how people think about success, and experience disability and illness, especially as they age? While many Japanese workplaces have mandatory retirement, people often return in a more limited role, earning less money to supplement their pensions but often working in valued social roles. Do people in more collectivist cultures see greater value in people playing different roles across the life course as their abilities change?

Fieldnotes

One of my earliest Japanese collaborators was a medical doctor specializing in women's health. He provided me with a room in his clinic in which to conduct interviews and introduced me to several participants. When he was diagnosed with Parkinson's disease, and was no longer able to run his clinic, he transitioned to providing what he called "kokoro no *care." The Japanese word* kokoro *can be translated as heart, mind, and/or spirit, and "care" is the English word care. Thus* "kokoro no *care" roughly means care of the heart/mind/spirit. As his ability to talk declined, he found that his ability to listen closely to people's concerns allowed him to practice an "empathy approach" to healing. This enabled him to continue to play an important societal role, providing an important service in the form of listening, and supporting people's holistic (integrated heart-mind-spirit) healing.*

To what extent might Japan's Eastern, collectivist emphasis on relationships and social connection influence experiences of aging and menopause? In Lock's research,[4] people claimed that women who were *isogashī* (busy) were less likely to be troubled by *kōnenki* problems. As a WEIRDo, I initially thought this seemed potentially sexist and even patronizing, suggesting that women with problems might just need to find more things to do. I first thought that perhaps Japanese women who were busy did not have the opportunity to complain about their symptoms, or perhaps even notice them? But years later, I found myself wondering, might it matter what they are "busy" with?

Fieldnotes

Japanese elders gather every morning in parks throughout Japan doing rajio taisō (radio calisthenics or exercise). *I have joined in to stretch and warm up before going jogging and noticed the camaraderie and connection between people who clearly have been doing it for some time, but also how welcoming they were to newcomers like me. These regular social and physical connections may be one factor contributing to longevity among Japanese. The social connection may be more important than the physical connection, as evidenced by a comment a physician made about elders who commonly came to his clinic, "many elders regularly come to the clinic. When they don't show up, we worry they might be unwell." He was implying that they came to the clinic more for social interaction, not because there was anything particularly ailing them.*

While it is easy for a Westerner to dismiss the idea that being busy might be an effective strategy compared to medical interventions for aging-related symptoms, a deeper exploration might shed light on complex biocultural processes that influence the experience of and attitudes toward health and aging. Could being busy, especially if it involves physical activity and social-emotional connections with others, not only change the brain but also lead to decreased stress and inflammation over the life course? Could a cultural factor of "being busy" influence proximate biology, and/or influence psychological perspectives on proximate biology and perception of associated symptoms? Evidence from functional magnetic resonance imaging shows that cultural values of individualism and collectivism can influence neural representations of self.[5] So cultural context does influence how our brains work!

Could Japanese collectivist culture and being busy with social connections (e.g., caring for children or elders) provide perspective and purpose, thereby influencing how people are bothered, and whether their symptoms interfere with their lives? Significantly, dependence on others (and obligations to respond) is so important in Japan, that it is marked linguistically with its own verb, *amaeru*. Doi defines the verb *amaeru* as "to depend and presume upon another's benevolence" much like young children depend on their parents.[6] Or at the end of life, how parents might depend on their children. Could elder dependence on, and connection to, younger people be one of the keys to healthy aging in Japan? Many people have experienced an increased ability to work or play through physical discomfort and pain. Activities that involve motivation and connection may mitigate mild to moderate symptoms and allow for greater activity, creating a positive feedback loop, with implications for aging. Could Japanese culture provide an example of how ultimate and proximate biocultural factors are intimately intertwined, and highlight how important it is to address social isolation as we age?

11.3 Cultural Attitudes and Expectations and Medicalization

Cross-cultural differences in attitudes toward aging and reproduction exist and may influence whether they are viewed negatively as a loss or positively as a relief or opportunity. Factors such as roles and status related to gender and age as well as medicalization, influence cross-cultural and within-cultural variation. Medicalization is an ultimate cultural process by which human conditions are defined and treated as medical conditions, and become the subject of medical study, diagnosis, prevention, and/or treatment.

Different cultures label different conditions as medical and in need of treatment. Medical conditions that appear unique to certain cultures are called "culture-bound syndromes" and include *susto* (illness attributed to terrifying event causing soul to leave the body leading to distress) in Latin American populations,[7] and imbalance of the autonomic nervous system (*jiritsu shinkei shitchōshō*) in Japan. Since Western medicine is hegemonic in the world, symptoms found in non-Western cultures are often considered to be culture-bound. But WEIRD cultures also have potential "culture-bound syndromes," possibly including some of the symptoms of aging and menopause, particularly when considered from an evolutionary perspective (see Part III). Even pregnancy, arguably quite a "natural" process, has been medicalized to varying degrees in different cultures.

Fieldnotes

I gave birth twice in Japan. The first time I was surprised to learn that my health insurance would not cover it because pregnancy and childbirth were not considered health problems. Japan has a very good system of paying for preventive screenings, so all my prenatal exams were covered. I received a coupon book for all my prenatal visits, as well as a maternal-child health handbook that tracks the mother's health throughout pregnancy and the child's health from birth to age six. But if health problems or complications arise, then they are covered by health insurance. My "natural" births without medical intervention were covered by the government's childbirth allowance. If I had required medical care (e.g., C-section), that would have been covered by my universal national health insurance.

Could the much better outcomes for maternal and child health (including much lower rates of maternal mortality and infant mortality) in Japan compared to the United States be influenced in part, perhaps ironically, because of Japan's lesser medicalization of the "natural" process of pregnancy and birth?

Culture influences where the lines are drawn between "normal" vs. "medically problematic" (in need of medical interventions). In the United States, temporary pain is often treated regardless of long-term costs, not just in childbirth (with use of epidurals potentially leading to more invasive procedures with higher risks), but also following surgeries with the idea that "pain is the 5th vital sign," a concept which has been increasingly abandoned due to unintended consequences.[8] Cultural attitudes toward pain (almost zero-tolerance in the United States compared to enduring pain and difficulty as embodied in the concept of *gaman* in Japan) may have potentially devastating consequences on long-term health due to use of highly addictive pain killers in the United States.[8]

"Culture-bound syndromes" highlight the culturally dependent nature of many health conditions, and how cultural context is critical to fully understand many health conditions. Non-WEIRD cultures might see WEIRD culture-bound syndromes in the medicalization, labeling, and treatment of normal grief that continues beyond some culturally defined cutoff as depressive disorder; existential angst as anxiety; premenstrual symptoms of depression as premenstrual dysphoric disorder (PMDD); and difficulties with attention and inability to sit still for many hours in school and work settings as attention deficit hyperactivity disorder (ADHD).

Many symptoms might be influenced by evolutionarily abnormal environments and behavioral expectations that exist in WEIRD cultures. This is not to say that such feelings and behaviors never reach a threshold of benefiting from medical intervention, particularly in certain environments. But society and culture often influence where thresholds are set. These are biocultural phenomena in that the culture may lead to varying biology (e.g., more frequent menstrual cycles), which in turn lead to symptoms, or culture may lead to expectations about behavior (sitting still for hours on end in front of computers) that depart from our evolutionary history and impact physiology. Thus, although Part IV focuses on ultimate cultural factors, biology is also implicated. Examining medicalization, particularly cross-culturally, allows us to question where the lines are drawn, and with what consequences for experiences of aging and menopause.

Aging as a medical condition has only recently been formalized in the West. The field of geriatrics only formally began in the United Kingdom in the 1950s and in the United States in the 1970s. It has become a growing medical specialty, with an entire branch of the US National Institutes of Health dedicated to it: the National Institute on Aging (NIA) founded in 1974. Medicalization often leads to proximate interventions, such as anti-aging treatments for skin (e.g., botox), and calcium and vitamin D supplements and drug therapy for conditions such as osteopenia and osteoporosis that are more than skin deep. Anti-aging medicine has even gained prominence in Japan, perhaps somewhat ironically, as it is the country with the longest life expectancy, and thus might seem to need "anti-aging" medical interventions less than many other countries!

By focusing on medical interventions for aging and menopause, what might we be ignoring? What might we learn if we look "upstream" to examine factors influencing differential experiences of aging and menopause throughout the world?

11.4 Social Determinants of Health (SDOH): Beyond Genes and Lifestyle

Why might some populations (defined as members of geographical, racial/ethnic, or social class groups) on average live longer, healthier lives than others? Variation in life expectancy is explained, less by genes than by proximate risk factors and behaviors, such as diet and exercise (with indicators such as body mass index, blood sugar, and blood pressure). But how many of these risk factors are ultimately attributable to individuals or under their control, as opposed to "upstream" ultimate factors that constrain their choices?

SDOH are upstream factors defined as the "conditions in the environments where people are born, live, learn, work, play, worship and age that affect a wide range of health, functioning, and quality-of-life outcomes and risks."[9] They are often divided into five domains: 1. education access and quality, 2. healthcare access and quality, 3. neighborhood and built environment, 4. social and community context, and 5. economic stability, all of which intersect with discrimination and inequities based on minority membership. From a biocultural perspective, it is important to consider how these cultural factors become embodied and lead to differences in health experiences and outcomes.

11.4.1 Searching Upstream for Ultimate Causes (and Solutions)

There is a well-known parable in public health that begins with a man and woman fishing along a river. Suddenly, they see a person floating down the river, struggling to stay alive. The fisherfolk rescue the person from the river. Then another person comes and must be saved. This happens all afternoon and the fisherfolk get very tired from pulling people from the river. Eventually they think, "We need to go upstream and find out why so many people are falling in the river." When they arrive, they discover people being drawn to the edge of a cliff to look at the river. There are no guardrails or safety supports, so many of them fall in. The fisherfolk work with the local community to build a fence to keep people from falling in the river. Occasionally some still fall, but many fewer than before. This story illustrates how investing upstream in prevention may be more effective than downstream interventions and treatment, capturing the essence of public health approaches. Figure 11.2 applies this upstream/downstream parable to aging and menopause.

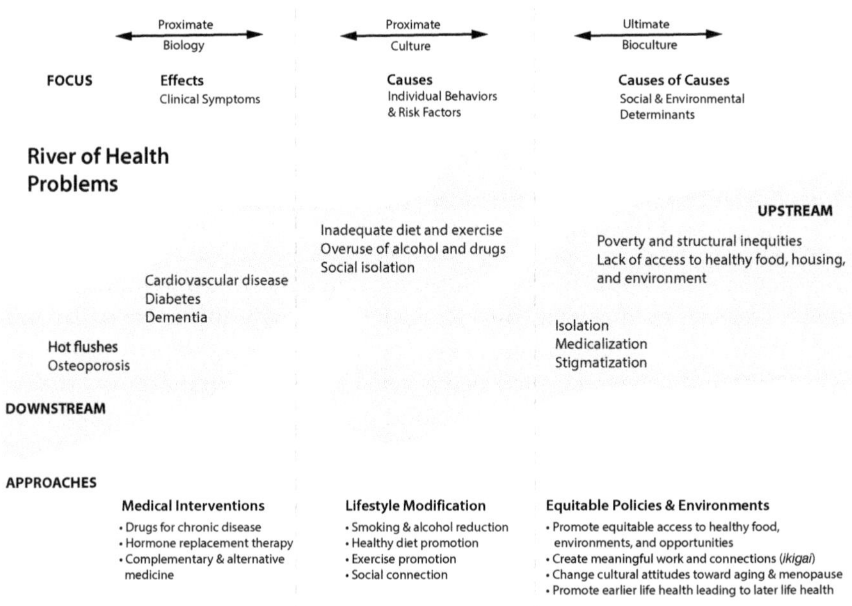

FIGURE 11.2 Schematic of the public health parable of the river, applied to aging and menopause. The river depicts upstream problems of poverty and medicalization leading to downstream aging-related diseases and menopausal symptoms. The focus downstream is on effects (e.g., clinical symptoms), while upstream, the focus is on causes and causes of causes (e.g., social and environmental determinants of health). Approaches range from downstream medical interventions, to mid-stream health behavior promotion and upstream policies. Moving from downstream to upstream (paralleling the focus of biomedicine to public health), these three areas of the river map to Proximate Biology (I), Proximate Culture (II), and Ultimate Culture (IV) quadrants.

In many industrialized countries, medical spending in the last few years of life is disproportionately higher than throughout much of the preceding life stages, likely due to costs of chronic conditions.[10] Chronic conditions constitute downstream problems. Public health recognizes that many lifestyle factors are out of people's control and looks upstream to social and environmental determinants of health to discover why people are falling into the river in the first place, examining systemic societal structures that create community environmental conditions that support or prevent health.

SDOH are shaped by the distribution of money, power, and resources within and between communities at different scales and contribute to health inequities throughout the world. The words "inequalities" and "inequities" often seem interchangeable, but there are important distinctions. The word "inequalities" suggests lack of equality without judgment about the ethics of difference. In contrast, the term "inequities" implies unfair and avoidable differences. Importantly, these words imply different solutions. Imagine three people of different heights, trying to pick an apple from a tall apple tree. The tallest can stand on tiptoes and reach an apple, but the shortest person cannot. Giving them all a similar size ladder would be *equal* treatment – and while it might allow the shortest person to reach the lowest apples, it would allow the tallest person to reach those and more, and thus would not be considered equitable. Giving different size ladders to each person, according to their height so that they have similar access to the resources of the apple tree would be an *equitable* intervention. Additionally, *justice* means that barriers are removed so that the need for accommodations is reduced. "Apples" could be job opportunities and "ladders" could be educational opportunities. Often those with the most resources have more tools (e.g., education) enabling them to access more resources, thereby exacerbating inequities.

A useful thought experiment when dealing with challenging problems (e.g., climate change, and inequities of income and health) is to ponder ways to exacerbate them. Often this generates more ideas as it can be easier to brainstorm ways to make things worse than to think of ways to solve complex problems. And then, by thinking of the opposite action, we sometimes discover potential solutions. For example, people have little difficulty brainstorming answers to the question, "how can we exacerbate educational inequities?" Answers include: Make education really expensive so poor people can't afford it at all, end subsidized school meal programs, pay teachers in some schools poorly so nobody wants the job, place ridiculous expectations on teachers (to fix problems at home, as well as behavioral problems in the classroom, and improve test scores year after year) so that turnover is high, and don't value teaching so it is considered a low-status occupation. Then solutions (often the opposite of the approaches to exacerbate the problems) seem clearer.

Another way to gain insight into how SDOH contribute to health inequities is to conduct cross-cultural comparative analyses to observe what other countries and cultures are doing. Japan has much better health outcomes than the United States, and while this may be in part due to genes and lifestyle factors, Japan often seems to have policies that minimize inequities, rather than exacerbate them, in numerous SDOH domains.

11.4.2 Japan and SDOH

Long-standing systemic health, social, and economic inequities have put some people at increased risk of getting disease. While Whites are often the default comparator when looking at inequities, particularly in the United States, globally they do not have the longest life expectancy or best health outcomes. The United States spends almost three times as much per capita on healthcare compared to Japan,[11] but Japan has better outcomes such as (healthy) life expectancy at birth and at age 60.[12]

Culture influences what people expect to receive in exchange for paying taxes (e.g., universal healthcare) and can influence behavior. For example, consider parental leave and expectations and behaviors related to breastfeeding. When I lived in Japan, the cultural ideal of breastfeeding for one year coincided, perhaps not coincidentally, with the maximum paid maternity leave of one year (paid at about 30%). This highlights how work culture may influence behavior and later life health risks, since breastfeeding may have effects not only on children's long-term health but also on women's later life health and aging, in part through effects on breast cancer rates which are inversely related to breastfeeding duration.[13] Next, we examine key differences between Japanese and WEIRD (e.g., US) cultures along several SDOH domains in turn.

11.4.2.1 Education Access and Quality

In many WEIRD countries, educational attainment is strongly correlated with health and thus education access strongly influences health inequities. In the United States, schools are funded primarily by local property taxes and curriculum is subject to local school boards. In contrast, in Japan, education curricula are more standardized across the country, and there is less variability in teacher pay (they are civil servants) and resources according to the local tax base. Furthermore, access to elite higher education is not as dependent on socioeconomic status as it is in the United States, and may be somewhat inversely related, as Kyoto and Tokyo universities are considered to be two of the top institutions and are public, not private, and the cost to students is relatively low (less than a tenth of the tuition at elite US private universities, and less than in-state tuition at many US state universities). Additionally in Japan, in the case of disadvantaged subpopulations (e.g., single mothers), there are programs that exist to provide free daycare as well as university tuition to women who choose to enroll in programs leading to high-demand occupations such as nursing. Such Japanese programs might help break the cycle of poverty that many US single mothers face in trying to study and cover childcare costs.

11.4.2.2 Healthcare Access and Quality

Japan has sliding scale health insurance premiums for participation in a national health system that provides equal access to providers and services. While the Affordable Care Act extended coverage to many previously uninsured individuals in the United States, it is still not universal healthcare, which all other industrialized countries have. In the United States, a medical emergency can lead to bankruptcy for people without adequate insurance, or delays going to the hospital and thus sometimes death.

Fieldnotes

As a graduate student living in Japan with limited income, my monthly health insurance premiums were 1/30th of the premiums of a friend who worked in IT at a bank making an upper middle class salary. This suggested a considerable sliding scale. People could go anywhere and get their care covered and only be responsible for a co-pay of 30% of the costs. Thirty percent may sound high to readers in the United States, but medical costs in Japan were low compared to the United States. For example, when my husband had chest pain, an angiogram, and a stent inserted during a five-day hospital stay, the total cost was a fraction of what it would have been in the United States. And then a Japanese friend informed me that we could apply for government reimbursement for most of the (still considerably lower) out-of-pocket expenses to minimize the economic burden of health crises. These reimbursements are available on a sliding scale according to income, but the scale slides quite high.

Importantly, inequities can be created not only by flat rate costs (e.g., health insurance premiums), but also by fixed percentage benefits (e.g., retirement contribution matches). Thus, in much of the United States, while absolute retirement benefits are sliding scale (favoring high-income earners), health insurance is not (penalizing poorer income earners), thus exacerbating economic inequities, a hallmark of many US institutions. What if the "rules" were reversed, and people paid sliding scale health insurance premiums (as in Japan), and retirement contributions from employers were a fixed amount per year? This would serve to decrease income inequities considerably.

11.4.2.3 Neighborhood and Built Environment

The built environment influences air quality, exercise opportunities, personal safety, and social connection. Japan has few guns and low rates of violence against strangers. There are many parks and excellent public transit in major urban areas, where increasing numbers of Japanese live, encouraging people to walk. In rural areas, many aging people are still farming and thus benefit from exercise and healthy food.

Neighborhoods in Japan often have mixed income housing, where wealthy homes exist next door to relatively cheap apartments. This contrasts with the United States, where neighborhoods are much more segregated by socioeconomic status. In Japanese cities like Tokyo and Osaka, Japanese CEOs often eat lunch in the company cafeteria and take public transportation,[14] just as my Director-General did when I was at the Japanese National Institute of Health. Thus, people of varying socioeconomic levels intermingle throughout the day and perhaps importantly, all expect high standards for their public transit and other infrastructure such as parks. This contrasts with the United States, where wealthy people rarely take buses or subways, and parks in different parts of the city look very different depending on the socioeconomic levels of the people living in the immediate vicinity.

11.4.2.4 Social and Community Context

Living in Japan, I was surprised to see the numerous public locations available for social and community connection. In larger cities, *jidōkan* (children's house, a type of community center for children and caregivers) are commonplace, providing indoor spaces for babies, toddlers, and elementary aged kids to crawl, play with blocks, build extensive train tracks, and for parents to connect. Moving costs can be high in Japan and provide a deterrent to moving, which may serve to create stronger community ties. When people do move in Japan, they give gifts (usually useful items such as soap and towels) to introduce themselves to their neighbors, building community, which can contribute to health. Parks are widely used for exercise and picnics, as well as practicing musical instruments. Cycling home after work along the river in Kyoto I often heard mini-recitals. Community centers are common through Japan, and many participants suggested we conduct our interviews there. One participant utilized these community centers to organize support groups for new mothers, building community for women at a time in their life course when they may feel particularly isolated. These centers were free and open to anyone to use, and thus illustrate use of government funding to promote social connection and community.

11.4.2.5 Economic Stability

Perhaps the greatest test of commitment to equity occurs in challenging times. Since the economic bubble burst in the 1990s, Japan has been experiencing a significant recession, but yet still has a relatively strong social safety net. Life has gotten considerably harder for many, with the rise of *furītā* (freeter, anglicized; referring to un/underemployed non-regular wage earners) culture reflecting and contributing to economic uncertainty, perhaps delaying marriage and decreasing birth rates. Nevertheless, while experiencing protracted economic hardship, Japan has attempted to buffer and minimize inequities that in other countries might quickly spiral out of control.

While living in Japan, I heard stories of bosses cutting their own salaries before laying people off – something I have rarely heard of in the United States where the ratio of CEO to typical worker salaries (400-fold difference in 2021) seems to have increased dramatically in the last few decades.[15] When I joined the Japanese National Institute of Health, the executive director earned about four times my salary. In contrast, when I joined my US university in 2011, the president earned over 14 times my salary (not including the benefits such as a free house he received), notably more than the president of the United States. My children found this puzzling, and asked if the job of the university president was more important than that of the US president, who earned considerably less.

Taken together, these SDOH and policies likely have many downstream effects contributing to Japan's hallmark longevity and shed light on factors potentially influencing inequities in many populations. In Chapter 12, we examine ultimate cultural factors influencing healthy aging, with a particular focus on Japan.

References

1. Henrich J, Heine SJ, Norenzayan A. 2010. The weirdest people in the world? *Behav Brain Sci.* 33(2–3):61–83. https://doi.org/10.1017/S0140525X0999152X
2. Troisi A. 2020. Social stress and psychiatric disorders: Evolutionary reflections on debated questions. *Neurosci Biobehav Rev.* 116:461–469. https://doi.org/10.1016/j.neubiorev.2020.07.016
3. Hofstede G. 2011. Dimensionalizing cultures: The hofstede model in context. Online Read Psychol Cult. 2(1):1–26. https://doi.org/10.9707/2307-0919.1014
4. Lock M. 1993. *Encounters with Aging: Mythologies of menopause in Japan and North America.* Berkeley, CA: University of California Press.
5. Chiao JY, Harada T et al. 2009. Neural basis of individualistic and collectivistic views of self. *Hum Brain Mapp.* 30(9):2813–2820. https://doi.org/10.1002/hbm.20707
6. Doi T. 1973. *The Anatomy of Dependence.* Tokyo: Kodansha International.
7. Martínez-Radl FB, Hinton DE, Stangier U. 2023. Susto as a cultural conceptualization of distress: Existing research and aspects to consider for future investigations. *Transcult Psychiatry.* 60(4):690–702. https://doi.org/10.1177/13634615231163986
8. Levy N, Sturgess J, Mills P. 2018. "Pain as the fifth vital sign" and dependence on the "numerical pain scale" is being abandoned in the US: Why? *Br J Anaesth.* 120(3):435–438. https://doi.org/10.1016/j.bja.2017.11.098

9. Healthy People 2030. 2025. *Social Determinants of Health*. Available at: https://odphp.health.gov/healthypeople/priority-areas/social-determinants-health. Accessed 4 Feb, 2025.
10. French EB, McCauley J et al. 2017. End-of-life medical spending in last twelve months of life is lower than previously reported. *Health Aff.* 36(7):1211–1217. https://doi.org/10.1377/hlthaff.2017.0174
11. World Bank Group. 2024. *Current Health Expenditure per Capita*. Available at: https://data.worldbank.org/indicator/SH.XPD.CHEX.PC.CD. Accessed 4 Feb, 2025.
12. WHO. 2020. *Life Expectancy and Health Life Expectancy Data by Country*. Available at: https://apps.who.int/gho/data/node.main.688. Accessed 4 Feb, 2025.
13. Stordal B. 2023. Breastfeeding reduces the risk of breast cancer: A call for action in high-income countries with low rates of breastfeeding. *Cancer Med.* 12(4): 4616–4625. https://doi.org/10.1002/cam4.5288
14. Koike Y. 2015. *Why Inequality Is Different in Japan*. Available at: https://www.weforum.org/stories/2015/03/why-inequality-is-different-in-japan/#:~:text=Japan%20has%20a%20high%20income,a%20significant%20driver%20of%20inequality. Accessed 4 Feb, 2025.
15. Bivens J, Kandra J. 2022. *CEO Pay Has Skyrocketed 1460% since 1978*. Available at: https://www.epi.org/publication/ceo-pay-in-2021/#:~:text=CEO%20pay%20has%20skyrocketed%201%2C460,in%202021%20%7C%20Economic%20Policy%20Institute. Accessed 4 Feb, 2025.

12

ULTIMATE CULTURE OF AGING

If we are interested in aging, and especially the secrets to healthy aging, Japan provides a particularly informative case study because it is leading the aging society "revolution." Life expectancy is frequently used as an indicator of population health, and by that measure, Japan has led the world in health for many years. But the existence of long-lived people alone does not make an aging society. Declining birth rates and low levels of immigration are also key to the demographic shifts that are occurring to cause the Japanese population to age so rapidly.

Much alarm has been raised about aging populations worldwide. The upstream causes and downstream impacts of these changes in population structure are historically unprecedented. People, and particularly governments, often focus on who will support and care for these elders, but less about what elders have done and can do for their communities, and how they can stay healthy, thereby extending not just life expectancy but healthy life expectancy. Comparative analyses of hot spots of longevity (i.e., Blue Zones) may provide important clues. Not all older people need care or financial support, and many may give more to society than they take. A reframing of the costs and benefits of aging societies, and the roles of aging individuals might be needed.

12.1 Japan's Aging Society, in Global Context

12.1.1 Causes, Concerns, and Consequences of Japan's Aging Society

While the earth's human population continues to grow, in the last 75 years, the population structures have shifted from primarily young to increasingly

DOI: 10.4324/9781003644521-16

old. In 1950, there were 2.5 billion people on the planet, and by the 2020s, the population reached 8 billion.[1] During this time, the global age distribution changed dramatically. In 1950, the age distribution approximated a pyramid, with many more younger people at the base of the pyramid compared to numbers of older people at the top of the age pyramid. The "population pyramid" is a classic bar graph of the distribution of a population, usually divided into age groups and males vs. females (see: population pyramids simulator).[2] When populations are growing, the graph typically takes the shape of a pyramid. But when populations are shrinking, due to decreases in birth rates and low immigration, and extensions of life expectancy, the population structure shifts from a pyramid to resemble a rectangle or box (see Gapminder[3] for interactive resources). A demographic structure characterized by more older people than young children is unprecedented in human history and poses challenges to existing health and social systems. But whether elders impose net costs or offer net benefits depends on ultimate cultural structures and population health.

In Japan, the population is aging faster than in most other countries. What are the implications of this dramatic demographic shift, particularly the relative proportions of people of working ages vs. those in retirement? Economically, how do retirement and pension policies impact who supports whom and how?

Population pyramids capture national-level phenomena, but these changes play out on household levels in roles and expectations. Many study participants talked about taking care of in-laws, but most said they did not want to be a burden to their daughters-in-law. In this sense, many of the women felt they were part of a sandwich generation – of women who were still expected to care for their parents-in-law while also caring for their children and grandchildren –but many did not want or expect to receive similar care when they aged.

Fieldnotes

Conducting an interview in a slightly darkened room in a community center, I recall how intense the woman was when she told me she had not wanted to be responsible for taking care of giri no oya *(parents-in-law), and so she committed to not marrying a first-born son. Particularly in more rural areas of Japan, a daughter-in-law's work can be quite onerous, involving taking care of not only parents-in-law, but their family farm as well. Several years after she married the second-born son, the first-born son passed away, and her husband was expected to move in with his parents and fulfill the first-born son's role. This made her the wife of a de facto first-born son, and she had to step into the role of daughter-in-law that she had worked so hard to avoid. She told me that this caused her to become severely depressed, and although she now felt better, she told me this with tears in her eyes, seemingly mourning the loss of her own vision for her life that she had worked so hard to achieve.*

Household structures are shifting from multigenerational homes to nuclear families albeit with extended family connections outside the structure of the home. Alongside these changes, family size has also been decreasing. The Japanese population is shrinking, largely due to Japanese having fewer children. This is occurring, not through government mandates to limit family size (as was done in China), but through societal factors that influence individual choices about marriage and children, such as delaying or avoiding marriage altogether and having fewer or no children. And these factors are the focus of many governmental initiatives.

Ultimately, these demographic shifts resulting from individual reproductive decisions create significant societal, economic, and government challenges, such as changing government financial balance sheets (incoming revenue and outgoing expenditures). Concerns about global population impacts on planetary sustainability are ongoing but some people are becoming increasingly concerned about underpopulation: not in terms of impact on the planet, but in terms of socioeconomic sustainability. By 2070, Japan's population is expected to fall by a third from its 2008 peak of 128 million, squeezing the economy and increasing national debt.[4] Projections in Japan suggest that the proportion of people 65 or over will grow to make up 40% by the year 2060, when there will be just 1.3 workers for every senior citizen, increasing the burden on the working-age population.[5] Contrast this with 1990, at the height of the Japanese economic bubble, when the ratio was about six people of working age for every senior citizen.[5] What are the implications for society, the economy, and people of working age who will be expected to support the elderly directly and/or indirectly? And how will the impacts be distributed throughout society?

12.1.2 *Variation in Aging within Japan*

Although they usually show age and gender, population pyramids or boxes combine multiple identities in the bars of the graph. Despite stereotypes, Japanese culture is not monolithic nor homogeneous. Women may live longer than men worldwide and in Japan, but Japan has done relatively poorly on reducing other aspects of the gender gap, such as political economic participation.[6] Broad brushstrokes suggest that Japan differs from WEIRD societies in many ways, but there is also considerable variation within Japanese society by factors such as age, gender, class, politics, religion, geography, caste, and ethnicity. The latter may surprise some readers, who may have heard that Japan was an ethnically homogeneous society, but the Japanese population includes indigenous populations such as the Ainu and Okinawans, and more recent immigrants to the archipelago such as Koreans, some of whom have resided in Japan for many generations. Japan also has populations belonging to lower castes or descent-based groups, such as *Burakumin* (short for *hisabetsu Burakumin*, people of discriminated villages).

Fieldnotes

I can't recall when I first heard the word Burakumin. *In Kyoto, after living with a host family for a few months, I decided to look for my own place. I did not have much budget for housing and was surprised to find a very cheap rental house in the NW corner of Kyoto, an area called Takagamine, advertised to foreigners. When I told some of my Japanese colleagues and friends that I planned to move there, their facial expressions suggested some surprise, but nobody explained why. I thought perhaps it was because it was on the edge of Kyoto, and thus farther from my university, or because they were surprised that as a foreigner, I had found someone willing to rent to me for a year. That may have all been true. But when I learned that Takagamine was also known to be a* Burakumin *area, I surmised that that fact may have played a role and might have been why the landlord was advertising to foreigners (something most Japanese landlords didn't like to do).*

It is hard to get many Japanese to talk about Burakumin, *and it seems to be a taboo word to many people.* Burakumin *literally means hamlet people, referring to laborers of impure occupations like butchers and undertakers. An address in Takagamine apparently could identify people as* Burakumin, *and since address of birthplace is recorded on family registries in Japan (previously often requested by employers but now restricted), it could be a means to discriminate. While employer-based discrimination may have decreased, lists circulate to allow people to prevent their children from marrying a person with* Burakumin *ancestry.*

When I first went to the field, I was influenced by the common misconception that Japan was relatively homogeneous, and so I did not ask about ethnicity or caste in my research. However, I did collect socioeconomic data since socioeconomic inequality in the United States strongly correlates with health inequalities. Japan has less income extremes than the United States: the top 1% of Japanese households earn about 1/5 of the top 1% of US households, and policies such as relatively high inheritance tax rates hinder generational wealth accumulation and transfer, thereby minimizing inequality.[7] More research is needed on how experiences of aging might differ within Japan among groups with different backgrounds, and how those backgrounds create different local or situated biologies within Japan.[8]

12.1.3 Interventions – Focus on Birth Rate and Immigration or Increased Health of Elders?

The Japanese government has had several responses to the demographic changes associated with an aging society. First the government has tried to increase birth rates, by increasing marriage and offering financial incentives to people to have children. Immigration is another approach, but one that has not been used much compared to other countries. And if population can't be increased by increasing birth rates or migration, then increasing later life roles for elders, as well as healthy life expectancy so they require fewer resources at the end of life and/or can contribute is another strategy for shifting the impact of the demographic changes.

The Japanese government has intervened by organizing and sponsoring matchmaking events and dating apps to encourage people to get married and have more babies.[9] And they subsidize child rearing costs, by providing monthly payments to families with children under a certain age.

Fieldnotes

When I lived in Japan, we received jidō teate *(monthly child payments) for each of our children. The value at the time was about 3% of the monthly median household income for a child under three, and half that (1.5%) for a child aged three to six. But anybody who has raised children knows that that does not come close to covering expenses (ranging from housing, food, and clothes, to transportation, education, and healthcare). While these payments may not provide adequate financial incentives for people to have children, they may at least offset some of the costs for those who decide to have them. Healthcare for children is free in many regions (e.g., 23 wards of Tokyo), with the remaining 30% after insurance pays being covered by local governments. Such differences between the United States and Japan have significant financial implications for raising families (even for my own when I returned to the United States and lost all these wonderful socioeconomic supports). But these benefits don't seem to have substantially impacted the decreasing birth rates.*

Given economic concerns about a decreased labor force supporting an increasingly elderly society, it makes sense for governments to intervene financially to incentivize and support having children. Covering medical costs is wonderful, but most children spend more time well than ill, and the majority of costs associated with non-medical expenses are not covered. Money is not the only constraint, as children require considerable time investments. The money provided is inadequate to compensate for (primarily) women's unpaid labor in childcare and housework. Work-life balance challenges can be high in Japan, especially for women, where they still are expected to do the majority of child rearing and housework.[10]

Increasing immigration is one option to solve Japan's demographic challenges, but Japan has been relatively reluctant to allow immigration, despite their decreasing population. When I lived in Japan, most immigrants could only become permanent residents after living in Japan for ten years (and working for at least five years), which was one of the longest requirements in the world. Although Japan has begun to relax their immigration requirements, compared to many countries, Japan remains hard to immigrate to and increasingly less desirable with the ongoing economic challenges.

If strategies to increase the population (especially of workers) have been unsuccessful, then strategies to increase work being done by elders might be another approach to consider.

12.1.4 Reframing Costs of Aging Society

Do attitudes and expectations about aging change with shifting demographics? The decreasing numbers of workers supporting every senior via pension payments may cause concern among younger generations, who not only fear that they will not be able to receive pensions, but that more of their income will be taxed to support previous generations, while they or their spouses will be expected to provide more care for elders. This seems like a dire situation, which is playing out in many countries around the globe. Yet not all workers are retiring when they reach retirement age.

The "old-age dependency ratio" measures the number of those aged above 65 years (currently defined by the World Bank as old age) as a share of those between 15 and 64 years (currently defined as working age). In other words, this ratio tells us how many potentially retired people a potential worker has to sustain. However, this ratio does not account for changes in healthy life expectancy as well as behavioral and economic changes. With changing roles, and better health and longer life expectancy, previous calculations and models on old-age dependency may no longer apply.[11] In many countries, more people over age 65 are working, often even part-time, partly due to better health, greater need due to economic realities of declining or non-existent pensions, or phased and other retirement options. Also, more women are joining the labor force in later life after raising children. These changes, as well as retirement policy changes, have brought more people into the labor market, and they often stay longer.

Because of these changes, an alternative ratio may provide more insight and better data for planning: the "adult dependency ratio," defined as those people who are economically inactive vs. active in the entire adult population, aged 15 and older. People may be inactive in the labor force for a variety of reasons – to care for children or elders (typically unpaid and unmeasured labor), or because of disability, and they may go in and out of active involvement in the labor force over their lives. This adult dependency ratio may paint a very different picture of the future. Consider Germany, which like Japan, also has a low birth rate. In the 1990s, ten German workers supported approximately seven inactive adult people (including many women), but this is projected to increase only slightly to about eight inactive adults per ten workers, and then decrease around 2050. In many countries, the adult dependency ratio appears more stable than the old-age dependency ratio.[11] So perhaps some of the concerns about aging societies are unwarranted, and governments may not need to get involved in people's personal lives as much as they think (i.e., trying to get people married and to have babies).

In Japan, despite an official mandatory retirement age, many people continue working in some capacity, often at a reduced salary, supplemented by their pensions. If older people continue working, are they taking jobs from young people? Possibly, but in countries with fewer young people that might be less of an issue. Also, the nature of jobs is changing. Many older people may become entrepreneurs or consultants after they retire, or volunteer doing things that other people might need to be paid for if it were not for their "unpaid labor." Social/political experiments with universal basic incomes for everyone, and more freelance and contract work are changing people's relationship to work (for better or worse, it is not yet entirely clear).

Why are countries so concerned about falling birth rates and aging populations? The primary reasons are probably the decreased working population relative to elders, and increased costs of pensions and healthcare. Yet for a holistic picture of impacts, we must consider ultimate cultural domains beyond economic issues, including social, environmental, and political (Table 12.1).

In many analyses of aging populations, productive (measured in currency) and reproductive (measured in reproductive success, see Part III) contributions predominate, and unpaid societal contributions are often overlooked. In evolutionary circles, there has been great attention paid to the grandmother hypothesis – the idea that women who outlive their reproductive potential can accrue evolutionary benefits by investing in their grandchildren. But these benefits depend largely on grandparents living near grandchildren, being healthy enough and having adequate time and money to care for them.

TABLE 12.1 Consequences, including costs and benefits, of aging populations by ultimate cultural domain

Ultimate cultural domain	*Costs*	*Benefits*
Economic	Increased pension and healthcare costs (relative to working people paying into systems)	Unpaid volunteer work by retired people
Social	Increased burden of eldercare, born by individuals/families as well as society more broadly	Elders care for young (grand)children; engage in community volunteer work
Environmental	Extension of high-consumption lifestyle	Elders may use more public transit (when available); may have more time to invest in environmental movements
Political	Increasing costs spent on elders (pensions and healthcare)	Elders may take on senior leadership roles

Therefore, whether elders impose a net cost on society or provide a net benefit very much depends on social structures and health status of individuals.

While many postmenopausal women may be past their *re*productive potential, they may continue to be productive, not just in terms of money, but also in terms of unpaid or under-paid but critical labor in their families and communities (e.g., caregiving and community leadership). In Japan, postmenopausal women may be caregivers not only for grandchildren, but also for their elderly relatives, or others' parents and grandchildren when they work as home helpers or daycare workers. They may also assume leadership roles in their professions and communities.

Fieldnotes

After I had my first child and I became a convert to baby-wearing with the traditional Japanese onbu *(baby carrier), a participant and friend founded an* onbu-no-kai *(baby carrier group), to support mothers with young babies and encourage them to carry them, breastfeed, etc. I attended the group's meetings and events whenever I was in town and felt like my foreigner/outsider status might lend some legitimacy to traditional Japanese child rearing practices (and counter Western influences), and I was more than happy to oblige.*

She was always doing some kind of community support activity, and often brought me along to help. Later, after a long career as a nurse midwife, at an age when many people retire, she became the director of nursing at a nursing home facility, transferring her skills from serving a population that was decreasing (pregnant women and babies) to one that was growing rapidly (elderly).

Just as this participant shifted her own professional focus from midwifery and babies to elders, many physicians were doing the same thing, perhaps influencing the medicalization of menopause. One participant told me that with falling birth rates in Japan, obstetricians had less work, so they were shifting toward caring for menopausal women, particularly as that demographic was growing. For these doctors, shifts from obstetrics to gynecology meant no more middle-of-the-night calls to deliver babies, or life-or-death decisions. This also corresponded to shifts in emphasis of women's roles, from women as reproducers to women as important caregivers, particularly of the elderly. Thus, there is a growing need for midlife women to remain healthy, productive members of society, often by providing care for elderly as well as children and grandchildren (including adult children, many of whom continue to live at home).

To increase the benefit to cost ratio of elders, societies need to keep them healthy. Most people are interested in aging "well," not just living "long" regardless of health status. In the last century, in WEIRD countries, life expectancy from age 45 has increased ten years, and from age 65 has increased five years. But not all that extra time is healthy. In fact, despite life expectancies around 75–80, estimates for healthy life expectancy in most European countries hover around age 65.[12] Roughly half of the additional years expected at age 65 are likely to have disability,[12] and this has implications for individual quality of life and societal-level policies and economics. While "downstream" medical interventions (Figure 11.2) effectively extend life expectancy, "upstream" interventions may be required to extend healthy life expectancy (additional years spent in good health without having activity limitations, assuming they started life without them). The next section examines aging in distinctive geographical regions called Blue Zones, which provide clues to ultimate cultural factors that influence healthy aging.

12.2 Blue Zones

Even though evolutionary theories suggest it will be difficult to "cure" aging, people have long searched for the secrets to living as long and healthily as possible. Cross-culturally, there is great variation in how well people seem to age. Many researchers have looked to Japan for the proverbial "fountain of youth" given the long-life expectancy and high numbers of centenarians in the population. Many have focused on one location, Okinawa, which has the highest proportion of centenarians in Japan. Several other longevity hot spots or "longevity oases" where people are believed to have the longest life expectancies on earth exist outside Japan, and cross-cultural comparisons can help us understand additional factors that might lead to differential health outcomes and provide solutions for the rest of us.

12.2.1 What Are Blue Zones?

To elucidate keys to longevity and healthy aging, researchers have focused on regions where people live longer than average, called "Blue Zones."[13] Several Blue Zones have been identified: Okinawa, Japan; Sardinia, Italy; Ikaria, Greece; Nicoya Peninsula, Costa Rica; and Loma Linda, United States. They are called Blue Zones because when Dan Buettner and his colleagues were searching for these areas, they drew blue circles around them on a map (and the name stuck). They studied these zones to find the "ingredients" associated with healthier and happier lives and longevity.

While "good genes" contribute only about 30% of the variation in life expectancy, the rest comes from environmental and lifestyle factors such as diet (e.g., Mediterranean, primarily plant-based), physical activity, and social connections, and may be what gives these populations an edge on longevity (Table 12.2).[14] While inhabitants of Blue Zones experience age-related decline in organ function, fewer people experience chronic diseases such as diabetes, suggesting that such diseases are not inevitable outcomes of aging.

Fieldnotes

Interestingly, my initial research contacts in Japan were influenced by a Blue Zone, but not the one that is in Japan (Okinawa). Despite the United States having arguably the best medical care in the world (or at least the most expensive), only one of the Blue Zones is in the United States. Researchers discovered that a group of Seventh Day Adventists in Loma Linda, California suffered from only a fraction of the diseases that commonly kill people in other parts of the United States and industrialized world.

When I first went to Japan to assess feasibility of my research project on midlife women's health and influence of soy on menopause, my initial contact who introduced me to my two primary collaborators in Kyoto and Fukushima, was a doctor who had trained at Loma Linda University. He was working at a Seventh Day Adventist hospital in Tokyo and shared my interest in soy and health. Years later, when invited to write a chapter on vegetarianism and longevity for a Japanese medical text on Anti-Aging Medicine,[16] *I relied heavily on literature on Seventh Day Adventists, which is both a religion and a lifestyle, with a heavily plant-based diet. Their guide to optimal health is summarized by the acronym NEWSTART: Nutrition, Exercise, Water, Sunlight, Temperance (balance, moderation), Air, Rest, and Trust in God.*

12.2.2 Blue Zones Highlights from Japan

Women in Okinawa have the longest lifespan in the world in part due to lower rates of cardiovascular disease and cancer. Much research has focused on their diet, characterized by low sodium, and high intake of bitter vegetables and herbs (containing antioxidants), seaweed, sweet potatoes, fish, tofu, and surprisingly pork (all parts including collagen-rich parts). They also practice *hara hachi bu* which refers to eating until your stomach is 80% full. This implies that Okinawans may have "discovered" a caloric restriction approach to slow aging without the benefit of laboratory data (see Part I). How do they achieve this caloric restriction in a healthy way that does not contribute to eating disorders? Eating more slowly, focusing on the food

TABLE 12.2 Blue Zone characteristics[14,15]

Location	*Country*	*Diet*	*Activity*	*Work & social*	*Meaning/ spiritual*
Okinawa (oldest women)	Japan	Soy-based foods *Hara hachi bu* (eating until you are 80% full)	Taichi	Farming community	Meditative Taichi *Ikigai* (life purpose)
Sardinia (oldest men)	Italy	Med Diet*		Farming	
Icaria	Greece	Med Diet*			
Nicoya	Costa Rica	Beans and corn tortillas	Physical labor		"Plan de vida" (reason to live)
Loma Linda	USA	Strict vegetarians		Tight-knit religious community	Seventh Day Adventist

* Med Diet: Mediterranean diet rich in olive oil, homegrown vegetables, red wine (But note that in 2023 the World Health Organization proclaimed that no level of alcohol is safe. The widespread idea of moderate alcohol intake being healthier than no alcohol intake derives in part from studies that lump heavy drinkers who quit with people who never drank, some due to health conditions).

(rather than TV or cell phones), and eating with others in certain settings can lead to slower eating and feeling content before overeating.

Caloric restriction as a means to extend life expectancy has been studied in laboratory animals by using undernutrition without malnutrition. Hypothesized mechanisms include decreased cellular free radicals and increased endogenous production of antioxidants. Yet, results from caloric restriction experiments in primates have been less clear than in other animals and may provide clues to ultimate cultural factors relevant to humans. In a Wisconsin lab, calorie-restricted monkeys lived longer than controls; yet in a Baltimore lab, no significant difference in survival was found.[17] Why the discrepancy? One possible explanation is that in the Wisconsin lab, controls ate like average Americans (as much as they wanted of a not particularly healthy diet, loaded with sucrose, and in which all foods were processed in the laboratory), while in the National Institute on Aging lab in Baltimore, controls were fed whole foods (more evolutionarily normal) and were given fixed portions, based on what they naturally ate before the study started.[17] Thus, even the Baltimore control animals had a diet that was "environmentally" restricted in calories, at least to a moderate extent. The Baltimore study suggests that minor restriction of calories may be just as effective as significant restriction. This has important implications for ultimate cultural policies.

Perhaps the apparent calorie restriction that occurs in Blue Zones such as Okinawa results from ultimate societal environments that provide evolutionarily normal calorie constraints that occurred in humans' historical environments. Such environments may still exist in many communities where people grow much of their own (nutritionally and fiber-dense) food. It could be that such diets are particularly good for healthy aging, and/or that contemporary diets of almost unlimited quantities of highly processed foods are particularly bad for healthy aging, or both.

Given the current environments in which many humans live (particularly in industrialized countries), where excess energy intake may coexist with undernutrition (e.g., of fiber, vitamins, minerals, antioxidants), humans may be more like primates in the Wisconsin lab. In many Blue Zones, people are eating what seems to be a calorie-restricted diet (or from an ultimate biological perspective, eating more like our ancestors likely did) in contrast to the rest of us who reside in contemporary environments that promote obesity and chronic disease.

Rather than a calorie-restricted diet, a better phrase might be calorie-constrained, or "evolutionarily guided" diet, and there are many ways to achieve this, such as by increasing nutrient and fiber to calorie ratio and decreasing processed foods. Results from these experiments, in combination with insights from the Blue Zones, suggest that changes to the ultimate cultural environment might have additional benefits not only for quantity of life (i.e., life expectancy) but also quality. Additionally, as effects of climate change become increasingly apparent, increasingly vegetarian diets may not only be healthier for people, but also for the planet.[18]

Dietary intake is only one part of the equation, and most Blue Zones' populations incorporate regular physical activity (often as part of farming and food procurement) into their daily lives. Furthermore, as in other Blue Zones, *ikigai* (life purpose; see Chapter 13) is important in Okinawa, as are social networks and connections.

Fieldnotes

One of my participants lived in a Buddhist temple in Kyoto, and when I arrived to interview her, I was thrilled to discover that I would get to enter the inner sanctum of a temple – since only a few rooms of a temple compound are usually open to the public. This particular temple wasn't one of the big tourist attractions in Kyoto, but it was a working temple nevertheless. I have always found the uncluttered, open rooms with sliding doors and tatami mats of Japanese temples to feel like an extension of the outside gardens. I feel a sense of peace in them, as I imagine many people do, and I assume that is intentional design.

In talking with my participant, who was the wife of the primary Buddhist priest of the temple, I realized that the light, peaceful feeling of the garden could only be achieved with incredibly hard labor of regular gardening and cleaning and maintenance – which is how she spent most of her time. Thus, the lightness of the temple that I experienced may have placed a heavy burden on her shoulders and body. Since I was interviewing her in part about her body, and its changes at midlife, I could not help but wonder about the impact of such a lifestyle on her health. Was she sacrificing her own health for the happiness and health of others? Do the benefits of living in a beautiful temple outweigh the physical costs of having to maintain it? Or might the physical exercise, as well as meaning and purpose in maintaining such a cultural heritage site, be good for her health?

12.2.3 Summary of Blue Zones

What's the secret formula for a healthy long life? The Blue Zones show that there is not a single place or single path to longevity and health, but there are some commonalities.[15] While genes can explain some of the variation in longevity, environment and lifestyle explain the majority.[19] People in Blue Zones primarily eat a diet consisting of fresh (unprocessed) foods low in simple carbohydrates, a majority of plants such as lots of beans and nuts, and little red meat. They practice fasting periodically and/or follow the 80% rule and ingest alcohol in moderation or not at all. They get daily physical activity, incorporated into their lives, not exercise spurts at the gym, burning five times as many calories as the rest of us in "non-exercise physical activity" and get adequate sleep. Elders in Japan, interviewed for another study I conducted on diet-related problems, often mentioned that ultimate cultural factors associated with modern industrialization and convenience (of both diet and activity) may be at least partly to blame, as human time and energy have been replaced with machines that rely on fossil fuels, impairing not only the health of humans but also the environment.

Living with multigenerational families may impose obligations, but also provides built-in social interactions that may benefit elders, as well as younger members of households and help to continue cultural traditions, including food ways and spiritual practices. Although their religious beliefs varied, having a strong belief in God was another trait inhabitants of Blue Zones had in common. People who had some form of spiritual practice four times a month gained on average 14 years to their life expectancy, and having a clear sense of life purpose adds about seven years to one's life compared to those who don't.[15] In Loma Linda, California, where Adventists comprise most of the population, the university graduates one of the highest percentages of registered dieticians and nutritionists in the world, connecting spiritual practices with diet.

Interestingly, Loma Linda appears to be maintaining their longevity better than the other Blue Zones. Obesity has become a problem among young people in Okinawa, and they no longer have the longest life expectancy in Japan (in part due to influence of US dietary culture, associated with the US military bases in Okinawa). Fast food has invaded Sardinia. So, the faith-based component (or cultural factors that ensure adherence to traditional lifestyles) may be a bigger factor helping people "stay on a path" to healthy aging.

Health guidance advises us to eat healthy foods, exercise, not smoke, etc. But to what extent are individual behaviors truly under our control, or dependent on upstream social determinants of health? It is important to note that race/ethnicity, gender, class, and socioeconomic status among other factors can influence people's access to important social determinants and resources (e.g., healthcare, healthy food, safe places to exercise, etc.). Having a purpose may provide incentives to engage in healthy behaviors. And social interaction may help achieve one's purpose in life (depending on what it is). Diet and physical activity are often considered individual responsibility, but some locales and cultures support healthier diets and activity levels more than others. On a more ultimate level, societal supports and barriers exist that influence and even determine proximate behaviors and exposures. For example, it is hard for people living in inner cities with food deserts without safe places to walk to implement the lessons from the Blue Zones. Cultural attitudes, values, and expectations, along with societal responsibility and policy interventions, influence equitable opportunities for implementing healthy aging advice. In Chapter 13, we turn to ultimate cultural factors involved in *kōnenki* in Japan.

References

1. UN. 2022. World Population Prospects 2022: Summary of Results. *Population Division*. Available at: https://www.un.org/development/desa/pd/content/World-Population-Prospects-2022. Accessed 4 Feb, 2025.
2. CPC. 2011. *Pyramids App: The Dynamics of Populations Structures*. Available at: https://pyramids.cpc.ac.uk/. Accessed 4 Feb, 2025.
3. Gapminder. 2025. *Gapminder Resources*. Available at: https://www.gapminder.org/resources/. Accessed 1 Feb 25, 2025.
4. 2023. *Japan's Population Projected to Fall to 87 Million in* 2070. Available at: https://www.nippon.com/en/japan-data/h01664/. Accessed 4 Feb, 2025.
5. 2014. *Graying Japan to Face Unprecedented Challenges*. Available at: https://www.nippon.com/en/features/h00079/. Accessed 4 Feb, 2025.
6. 2023. *Global Gender Gap Report 2023*. Available at: https://www.weforum.org/publications/global-gender-gap-report-2023/in-full/benchmarking-gender-gaps-2023/. Accessed 4 Feb, 2025.
7. Koike Y. 2015. *Why Inequality Is Different in Japan*. Available at: https://www.weforum.org/stories/2015/03/why-inequality-is-different-in-japan/#:~:text=Japan%20has%20a%20high%20income,a%20significant%20driver%20of%20inequality. Accessed 4 Feb, 2025.
8. Niewöhner J, Lock M. 2018. Situating local biologies: Anthropological perspectives on environment/human entanglements. *BioSocieties*. 13(4):681–697. https://doi.org/10.1057/s41292-017-0089-5
9. Craft L. 2023. *Japan's Government Will Play Romantic Matchmaker in a Bid to Shore Up the Birth Rate. Experts Say It Will Fail*. Available at: https://www.cbsnews.com/news/japan-birth-rate-population-decline-government-matchmaking-marriage-children/. Accessed 4 Feb, 2025.
10. Zhou Y. 2015. Career interruption of Japanese women: Why is it so hard to balance work and childcare? *Jpn Labor Rev*. 12(2):106–123. Available at: https://www.jil.go.jp/english/JLR/documents/2015/JLR46_zhou.pdf
11. Koettl J. 2015. *Did We Get the 'Old-age Dependency' of Aging Countries All Wrong?* Available at: https://www.brookings.edu/articles/did-we-get-the-old-age-dependency-of-aging-countries-all-wrong/. Accessed 4 Feb, 2025.
12. OECD. 2019. *Health at a Glance 2019: OECD Indicators*. Available at: https://doi.org/10.1787/4dd50c09-en. Accessed 4 Feb, 2025.
13. Poulain M, Pes GM et al. 2004. Identification of a geographic area characterized by extreme longevity in the Sardinia island. The AKEA study. *Exp Gerontol*. 39(9):1423–1429. https://doi.org/10.1016/j.exger.2004.06.016
14. Buettner D, Skemp S. 2016. Blue zones: Lessons from the world's longest lived. *Am J Lifestyle Med*. 10(5):318–321. https://doi.org/10.1177/1559827616637066
15. Buettner D. 2013. *How to Live to Be 100+*. Available at: https://www.youtube.com/watch?v=ff40YiMmVkU. Accessed 4 Feb, 2025.

16. Melby, M. K., 裕太 石川, and 国比古 富永. 2015. "ベジタリアンは長寿か一アンチエ イジング（抗加齢） 医学からの検証 (Vegetarianism and longevity 一 examination from the perspective of anti-aging medicine)." In アンチエイジング医学 の基礎と臨床 (Anti-aging medicine: foundation and clinical), edited by Japanese Board of Anti-Aging Medicine Specialists and Health Practitioners (3rd edition), pp. 243–244. Tokyo, Japan: Medical View Co.
17. Wade N. 2014. *Diet's Link to Longevity: After 2 Studies Diverge, a Search for Consensus*. 1 Apr, 2014.
18. BBC. 2021. *Climate Change: Do I Need to Stop Eating Meat?* Available at: https://www.bbc.com/news/explainers-59232599. Accessed 4 Feb, 2025.
19. Dato S, Rose G et al. 2017. The genetics of human longevity: An intricacy of genes, environment, culture and microbiome. *Mech Ageing Dev.* 165(Pt B):147–155. https://doi.org/10.1016/j.mad.2017.03.011

13

ULTIMATE CULTURE OF MENOPAUSE

Ultimate cultural analyses focus on cross-cultural and historical analyses to understand menopause and expectations about what will occur during this life transition. The distinction between illness, disease, and sickness is illustrative. Illness describes the subjective experience, a person's perceptions and their lived experience (including symptoms) (discussed in Part II). Disease refers to the biological phenomenon, usually the observable, objective clinical manifestation of (assumed) abnormality of physical function, classified by a health professional (discussed in Part I). Sickness encompasses social phenomena, roles or factors associated with illness. In Part IV, we examine how, particularly in the case of menopause, the ultimate cultural factor of medicalization functions in the process of turning symptoms of an illness into a disease, thereby influencing the societally defined sick role and associated societal response.

Different societies and individuals assess the extent to which a phenomenon or symptoms are over-medicalized or under-medicalized, and this varies across cultures and time. For example, homosexuality was listed as a mental disorder in the *Diagnostic and Statistical Manual of Mental Disorders* (*DSM*) and the International Classification of Diseases (ICD) until 1973, when it was removed, thereby de-pathologizing it.[1] The history of classification, medicalization, and various "treatments" of homosexuality demonstrate how cultural constructs change over time and across place. This is also true for menopause, as cross-national and cross-cultural studies as well as within-country cultural (ethnicity) comparisons have demonstrated.[2–4] Ultimate cultural analyses focus on cross-cultural and historical analyses to provide insight into these shifting lines.

Kōnenki in Japan can be understood on an ultimate cultural level against a cultural historical backdrop of Japan being an EEIRD (Eastern Educated Industrialized Rich Democratic) country. While Western medicalizing

DOI: 10.4324/9781003644521-17

influences have been increasing in Japan, a long history of Eastern philosophy and culture affects the form medicalization takes. Further, Japan's collectivist values influence how symptoms are perceived and responded to by individuals and those around them. More robust social safety nets and lower economic inequality (as measured by the GINI coefficient) in Japan may be related to the greater valuation placed on collectivism over individualism than in many Western countries. This cultural and historical backdrop provides context for Japanese attitudes toward women's roles, which influence attitudes toward, and experiences of, *kōnenki*.

13.1 Medicalization

Aging and menopause are biologically inevitable but culturally variable experiences and are influenced by biocultural factors. In cultures dominated by Western biomedicine, normal physical changes associated with menopause tend to be perceived as pathologies requiring both medical and cosmetic "help," perhaps because youth is so highly valued and aging is so medicalized.

The degree of medicalization can influence people's attitudes toward and experiences of "natural" life course events and processes. In Japan, views of *kōnenki* were influenced by German medical thought in the late 1800s,[5] and thus can be thought of as having experienced some degree of Western-influenced medicalization. But *kōnenki* (see Part II) is still viewed by many Japanese as a natural transition of aging, one that does not require medical intervention.

The more recent US-influenced medicalization of menopause in Japan was highlighted by a *Newsweek* Japanese special issue on women's health that was published during my fieldwork. The ethnicity of the woman featured on the front cover suggested that it was translated from the West, without much attempt to tailor it to local culture and local biologies, as it showcased hot flushes, but not symptoms of chilliness or *katakori* (shoulder stiffness) that are more prevalent in Japan. Unfortunately, the experiences of Western women at menopause have been assumed to be "universal" and extrapolated to the rest of the world in many cases. This assumption may blind us from recognizing human variation that exists cross-culturally, and prevent us from gaining a holistic biocultural understanding of the experience of this life period.

Fieldnotes

Some Japanese women were clearly influenced by the media, and on the lookout for what media told them were menopausal symptoms. I had told participants that my research project was a study of midlife women's health, and I purposefully did not mention menopause or kōnenki *because I didn't want to bias them toward reporting expected symptoms.*

Only at the final interview did I ask about kōnenki *and* heikei *(end of menstruation). The participant who was excited to finally think she was having a hot flush but then realized it was likely caused by the kerosene stove (see Chapter 7), demonstrated how the Western idea that menopause is inextricably linked to hot flushes had definitely permeated her consciousness. She was looking out for them, even though in my study most Japanese women experienced them less frequently than other vasomotor symptoms like chilliness.*

Medicalization influences the degree to which menopause is seen as a medical condition to be treated or more as a natural process like puberty. Interestingly, just as reproductive potential (e.g., pregnancy and ability to achieve or prevent pregnancy) has been medicalized, the end of reproductive potential (i.e., menopause) has also been increasingly medicalized. Following the adage "if the only tool you have is a hammer, everything begins to look like a nail," medicalization of menopause has also undoubtedly been influenced by the extraction and synthesis of estrogen and progesterone, as these hormones not only may provide relief for some menopausal symptoms, but also provide significant income for pharmaceutical companies and medical professionals, highlighting ultimate cultural factors of capitalism and economics. Medicalization of menopause may also be influenced by other ultimate social and demographic factors such as falling birthrates and aging populations that contribute to gynecologists' shifting focus from pregnancy to menopause (as discussed in Chapter 12).

To the extent that medicalization of menopause in Japan results from Western influences, it is useful to better understand the history of how menopause came to be perceived, in Western medicine, as a deficiency requiring medical intervention. Foxcroft argues that medicalization of menopause was not the result of deliberate policies so much as the outcome of increasing professionalization within medicine, growth of an affluent middle class, and an underlying misogyny that viewed women as natural invalids and their sexuality as something which required regulation.[6] There is a long history of attributing women's ill health and "undesirable" characteristics to women's reproductive organs and hormones. For example, terms like hysteria (derived from the Greek word *hystera*, which means "womb") reflect the belief that behaviors considered inappropriate for a woman were somehow caused by her uterus.[6] In the West, menopause as a subject began in 18th-century France, and gained traction in the 19th century, but ideas of women as pathological specimens were informed by ideas from the ancient (male) Greeks and Romans who believed males were the ideal form, while females were mutilated males and therefore abnormal, and that menopause was a "gateway to death."[6] Not all of these Western ideas have been exported so explicitly to Japan, and when they are, they often are incorporated in novel ways, such as when the idea of "hormone deficiency" becomes "hormone imbalance."

Even in the West, attitudes and approaches to menopause have changed over time. Throughout the 19th century, women were offered management of menopause. Next, the early 20th century ushered in endocrinology and hormonal drugs linked to women's sexuality, reproduction, and health. In the latter half of the 20th century, many doctors and women seemed to believe that replacing hormones would not only treat all signs of menopause, but also make women look and feel younger, and ward off diseases of aging. This belief persisted even though no well-designed long-term clinical trial of hormone treatment had been conducted until the Women's Health Initiative (WHI) study. To the surprise of many, in 2002, the WHI study showed that long-term hormone treatment may carry significant risks for some women, depending on timing and duration (see Part I). Following the WHI results, medicine returned to management approaches focused on treating short-term symptoms of the menopausal transition. And then, in the 2020s, the pendulum began to swing back, with hormonal treatments seeing a resurgence as part of what has been called a "menopause revolution"[7] that has reassessed risks and benefits of HRT, toward increasing numbers of providers, patients, and companies advocating for hormone treatment to prevent or slow many symptoms of menopause and aging.

Martin has shown us how language and metaphors frame and influence how we think, particularly in medical contexts.[8] The term "hormone *replacement* therapy" implies a hormonal "deficiency" condition due to ovarian "failure," because otherwise why would "replacement" be necessary? According to this view, menopause is a condition like thyroid deficiency or diabetes. If it is left untreated, it will lead to greater risk for many chronic diseases, a lower quality of life, and premature death. This view was the rationale for the widespread use of long-term hormone treatment for postmenopausal women in the late 20th century. According to the conventional wisdom of that time, if women's ovaries had failed, making them deficient, it made sense to replace the hormones. Is such an ultimate cultural framing somehow different and better in the 21st century?

Fieldnotes

Discussions of menopause and hormone treatments often assume that all women want life (and their bodies) to continue as before. But this assumption does not always hold true. While giving a presentation on my research to an audience of women, including several participants in my research project, I invited comments about how heikei *(end of menstruation that occurs at menopause) affected them. One woman said "I was relieved I could have sex with my husband without the worry of getting pregnant." Her friend, sitting next to her, chuckled and said "I was relieved I didn't have to have sex with my husband anymore." Two women, ostensibly experiencing the same proximate biological symptom (i.e., end of menstruation), with opposing emotional and behavioral responses! Same biology, but different cultural contexts and framing, with implications for each woman's life, marital relationship, and sexual behavior illustrating that context is very important!*

As Foxcroft says about women's attitudes toward sex at or after menopause, "it is up to the individual woman – and she may want to get on that [sexual] merry-go-round or she may want to get off it, and it's up to her."[9] Medicalization of menopause promises that with hormones, women can stay "forever young" or minimize many of the less desirable symptoms of aging. Such cultural pronouncements can be problematic, not recognizing variation in context and experience, and because they may influence women to make medical choices without fully informed consent about long-term risks and benefits.

The metaphors underlying ideas of deficiency imply an appropriate "normal" level of hormones as the ideal. Given variation in hormone levels on multiple scales, including time (ranging from daily and monthly to across the life course) and space (ranging from lower levels in natural fertility populations or those living at higher altitudes), how do we determine the appropriate hormone level to "replace" postmenopausally? Biologically, just as a shot of insulin does not replace the complex workings of the islets of the pancreas, an oral pill taken once a day containing estrogen (even if "bioidentical") and progesterone cannot replace the pulsatile release of hormones that exhibit circadian and longer rhythms. Nor do exogenous hormones necessarily address the changes in receptors and feedback relationships occurring throughout the hypothalamic-pituitary-gonadal axis and in other tissues (such as bones and heart). And, just as hormones are somewhat crude interventions, underlying metaphors are crude representations of biological complexity. In contrast to diabetes, where an insulin shot, however imperfect a replicate of pancreatic function, may save a person's life, menopause is not exactly a life-or-death situation – so might benefit from a more nuanced approach.

Given that Japanese women live longer and healthier lives than women in other populations and have much lower rates of hormone therapy (not only at menopause, but also earlier in life), perhaps there are alternative paths to healthy postmenopausal aging besides hormone therapy. The language used by many Japanese may provide clues as to why ultimate cultural attitudes of Japanese differ from those of many Western women. Could there be alternative cultural conceptualizations of the changes occurring during "The Change" of life that might broaden our perspectives and approaches?

13.2 How Women Frame *Kōnenki* (Climacteric)

13.2.1 How Women Talk about Kōnenki

Cross-cultural research, particularly on biomedical topics, involves challenges of translation not only of symptoms but also of ultimate cultural conceptualizations of health, the body, and the relationship to the larger physical and social environments. Meanings of symptoms are context-dependent, influenced by expectations and lived experiences in the physical and social environment.

In Japan, although menopause is often translated as *kōnenki* in Japanese, the term *kōnenki* is more accurately translated as climacteric, with menopause being translated as *heikei* (literally end of menstruation). However, in popular and even biomedical discourse, *kōnenki* generally substitutes for menopause. A related term, *kōnenki shōgai* could be translated as "climacteric problems or disorders."

If *kōnenki* is equivalent to menopause, how can we understand women who say they think they can avoid *kōnenki* altogether, or the existence of women who are clearly postmenopausal by biomedical criteria but say they never had *kōnenki*? Could *kōnenki* be more akin to illness, involving patient perceptions and expectations in social context? Could menopause be more akin to disease, involving observation of menstrual pattern changes and medicalized symptoms? Few Japanese women described *kōnenki* as a medical disease. Even among women whose definitions of *kōnenki* included *heikei* (end of menstruation), it was often one factor among many (often social/lifestyle) factors signifying *kōnenki*: "*Heikei*, children become independent, husband nears retirement, parents become elderly, etc.; psychological instability."

13.2.2 "Balance" of Hormones

Another departure from the biomedical paradigm was observed in my study of how women talk about hormonal changes. Most of the people who mentioned hormones actually said *horumon baransu* (hormone balance) specifically that *horumon baransu ga kuzureru* ("hormone balance" breaks down). The foreign loan words "hormone balance" seem to be an imported idea with a Japanese interpretation, since the focus in much of the West tends to be on decreases in hormones (hence "hormone replacement therapy"), yet Asian medicine (particularly Japanese *kampō*) tends to focus on balance.[10] This idea of balance as important to the conceptualization of *kōnenki* was reinforced by the post-*kōnenki* women who reported that their bodies had "stabilized." It also explains the relatively low use of HRT in Japan[11] compared to the United States. If the decreases in hormones are understood to cause symptoms of menopause, then treatments involving replacing those hormones makes sense. But if an imbalance in hormones is believed to be the cause, then it is unclear how adding hormones would help. Imbalance implies loss of stability and potential lows and highs of hormones. Interestingly, biological studies of hormones at the menopause suggest that spikes of estradiol do occur, lending credence to the Japanese understanding of hormonal changes during this time.

Not only do hormones become unbalanced, but so do the physical body and emotions as illustrated by these short explanations about *kōnenki* from my interviews:

- "Imbalance in physical strength and emotions (mind/heart/spirit)"
- "The transition period from adulthood to old age. The period when the body gradually declines; the body's internal balance breaks down."

...as well as family and lifestyle rhythms, seen from a larger social context, perhaps influenced by Japan's more collectivist culture:

- "Hormone balance breaks down. Domestic (family) cycles also change."
- "When lifestyle changes and physical condition changes occur."

Some women discussed *kōnenki* as a period of change in one's life, taking a life history or developmental perspective by comparing it to the changes of the adolescent transition:

- "Takes time to control one's emotions. Like adolescence, emotions are shaken up (disturbed)."

This last quote depicts *konenki* as a turning point in the life course. And the comparison to adolescence is important, as most people recognize a period of imbalance as girls go through puberty, and exogenous hormones are rarely used to intervene then. Similarly, at the opposite end of the reproductive lifespan, Japanese see symmetry in a temporary transitional period of imbalance in both hormones and emotions during *kōnenki*. In Japan, many women reported that *kōnenki* was one of several *kawarime* (turning points) in a woman's life.

While some women still characterize *kōnenki* as *jiritsu shinkei shitchōshō* (imbalance of the autonomic nervous system),[12] the concept of imbalance seems to have been retained, but the autonomic nervous system seems to have been largely replaced by hormones in more recent discussions of *kōnenki*. (Ironically just when the West may be finally expanding views of menopause to include the autonomic nervous system). This emphasis on hormones highlights more recent Western influences and the medicalization that has occurred in Japan, but its modification in the local Japanese context is shown by the emphasis on hormonal *balance* as opposed to hormonal *decrease*. The "hormone imbalance" occurring at *kōnenki* reflects a view of menopause as characterized by temporary hormonal changes, after which most women believed a new equilibrium is reached. Combined with *gaman* (described in more detail in Chapter 6), a cultural and philosophical attitude to enduring and navigating life's changes, many Japanese women seem to navigate *kōnenki* and the menopausal transition relatively easily and even flourish during it and beyond. Yet, some women do experience problems during *kōnenki*.

13.2.3 Kōnenki Shōgai *(Climacteric Problems)*

One of the most puzzling observations in my interviews was how some women said they thought they could avoid *kōnenki* completely. Lock's research findings that Japanese "women may be 10 years past their final menses and still not identify themselves as having gone through *kōnenki*" have been cited in the literature.[13] Just as with misquotes of Lock's work claiming that there is no [single] word for hot flush in Japan, subtleties of language provide important clues to what may really be going on. And just as Japanese has many words for hot flushes (see Part II), *kōnenki* may only be understood fully in reference to related terms such as *kōnenki shōgai*. Given the literal meaning of the Chinese characters comprising the word *kōnenki* (regeneration, years, season/stage), it was not surprising that the largest proportion of women (40.3%) said they thought *kōnenki* referred to an age period (*jiki*), when asked to define it in contrast to *kōnenki shōgai*, which refers to obstacles or difficulties. My observations aligned with Lock's observations that:

> Many of the women who state that *kōnenki* can be avoided entirely apparently have uppermost in their minds unpleasant symptoms that they associate with this stage of the life cycle but that they believe not everybody necessarily experiences. In contrast, women who dwell less on specific symptoms and focus instead on the more general signs of aging assume that everybody goes through *kōnenki*.
>
> *(p.8)*[5]

Lock[5] observed that different women use *kōnenki* differently, with some using it as a shorthand for *kōnenki shōgai*. Extending Lock's observations, I observed that individual women do not subscribe to only one definition, but that rather women's language and definitions of *kōnenki* change depending on their personal experiences and stage of *kōnenki* and the context. In a cautionary tale for interviewers, women often appear to conflate *kōnenki* and *kōnenki shōgai* when just asked about *kōnenki*, but when asked to contrast them, they "return" to the traditional view of *kōnenki* as *kawarime*, a turning point in the life course that everyone will reach:

- "*Kōnenki* is the *jiki* (time period [when you might begin to feel physical changes]); *kōnenki shōgai* is the appearance of actual symptoms."
- "*Kōnenki* is the period when there is an age-related turning point in a woman's life. I think *kōnenki shōgai* is the symptoms that differ depending on the person."
- "*Kōnenki* is the *jiki* that you feel yourself (from beginning to end); you feel *shōgai* in your body."

The woman in the last quote implies that only the person experiencing *kōnenki* can define its start and finish, explaining why *kōnenki* might not map perfectly to Western biomedical categories of menopause that rely primarily on menstrual patterns (see Part II). And just as Rosenberger observed for the term "autonomic nervous disorder" in the 1990s,[12] shifts in language use of *kōnenki* and related terms can only be understood in the context of discourse and relationships.

- "[*Kōnenki*] comes to everyone, so I think that *shōgai* can be overcome by *kimochi no mochiyō* (one's way of emotional management) and way of thinking."

Interestingly, this suggests that whether women experience *kōnenki shōgai* is partly influenced by how people respond to symptoms they perceive, which may be largely influenced by the social milieu, including expectations, as well as support and burdens they have in their lives. Thus, social determinants of health, and medicalization earlier in the life course (which affects exposure to exogenous hormones such as oral contraceptives), may also moderate and influence the physical experience of symptoms and in turn how those are interpreted and experienced.

13.2.4 Intra-Individual Variation in Definitions of Kōnenki

Definitions of *kōnenki* appear to have changed historically. The focus on "hormonal balance," derived from East Asian medicine and *kampō*, highlights the fact that while biomedicalization has occurred, concepts are interpreted and influenced by the local cultural context, which continues to evolve. In contrast to Lock's research, discussions of *kōnenki* during my field work did not include mention of housewives being more susceptible to *kōnenki shōgai*, or the need to be busy in order to avoid *kōnenki*, suggesting that the discourse among women has shifted from putting blame on the lifestyle and personality of individual women[12] to a more biological explanation of *kōnenki*. Nevertheless, women in both my study and Lock's study mentioned that "how one thought/felt about it" could influence the degree of *kōnenki shōgai*, suggesting the persistence of a belief in emotional/psychological control, perhaps related to the value placed on *gaman* (endurance).

The discourse on *kōnenki* has taken an increasingly biomedical tone over the past 20 years as reflected by one participant's highly medicalized view:

- "*Kōnenki shōgai* is *futei shūso* (general malaise, poor physical condition). Unable to carry on with daily life. Hormone injections, medicines are necessary. Psychological consultation is necessary."

Yet, the vast majority of participants in my study had less medicalized views and interviews demonstrated that a clear consensus does not exist. It appears that some women still hold a more traditional view, while others seem to give more biomedical explanations linked to hormones and menstrual periods and certain types of symptoms (popularized in the media), indicating considerable medicalization in the intervening decades between my study in the 21st century and Lock's research in the 1980s.

Most research on menopause, even cross-cultural work outside the West, has been heavily influenced by the biomedical paradigm. Seldom have women been asked to define the terms employed by researchers. But when they are asked, variation, not only inter-individual but also intra-individual, often emerges:

- "*Kōnenki* is a time period characterized by new circumstances such as children's independence, looking after elderly parents. *Kōnenki shōgai* results from changes in hormonal secretion, bodily changes, etc."

Another woman explained that if you yourself don't think *kōnenki* is hard, then you feel you do not have *kōnenki shōgai*. She first described *kōnenki* as:

- "Don't have the energy to do anything. Troublesome."

And then contrasted the two terms by saying that:

- "Even if you have *kōnenki*, if you don't have personal difficulties, then you feel like you don't have *kōnenki shōgai*."

Attitudes and expectations thus seem to play a crucial role in influencing how *kōnenki* is experienced. Another participant, a medical doctor who assigned her *kōnenki* status based on periods, explained *kōnenki* first primarily in terms of social factors and then gave a more biomedical explanation, all in the span of a 20-minute interview. This example demonstrates that the term *kōnenki* has a cornucopia of meanings from which women choose depending on the context (or question). Just as *kōnenki* is often characterized as a time when various events – hormonal changes of biological menopause, social and lifestyle changes, physical aging – overlap, definitions of *kōnenki* are multiple and varied in order to characterize this rich and complicated experience.

While Western biomedical influences are apparent in the increased mention of hormones by Japanese women when discussing *kōnenki*, the concept of balance, and distinctions between *kōnenki* and *kōnenki shōgai* suggest that Japanese women believe they have more control over how they navigate this transition and are not simply at the mercy of changing hormones. Importantly, it also raises questions about how we do research and ask questions, including being open to positive experiences which are not usually the focus of biomedical research which tends to view menopause more as a disease state.

13.2.5 Positive Changes and Opportunities

Women often described *kōnenki* as a time when social roles such as child rearing ended, presenting opportunities to find or create new identities.

- "I think it is when the serious things such as child rearing, children's education, etc. end, and the third stage of life begins. The first step toward enjoying the small remainder of one's life."

Positive aspects of the climacteric are often overlooked by researchers despite a preponderance of positive attitudes observed in studies of general population samples.[14,15] Compared to Western women, many Japanese women appear to focus less on symptoms and more on life transitions and changing social roles:

- "Women finish with their roles as mothers and can slowly begin to think of themselves."

The sense of potential embodied during *kōnenki* is captured in the phrase often used to describe subsequent plans: *jibun rashiku ikiru* (to live for/as oneself), which often becomes a goal once child rearing is finished.[16] For avoiding issues with *kōnenki*, Lock documented the importance of *isogashī* (busyness), which might correlate with whether a woman could "afford" to let menopausal symptoms bother her. Busyness could be a code word for lifestyle constraints that don't allow a woman to have time to notice or complain about symptoms (related to the concept of *gaman* discussed above). Alternatively, busyness could also indicate social connection and obligations that may buffer people from the challenges of aging, as suggested by the Blue Zones (see Chapter 12).

13.3 Expectations for *Kōnenki* and Beyond

To move beyond WEIRD perspectives and approaches, and to integrate the lived biocultural experiences of aging and menopause, this section explores ultimate cultural (historical) generational shifts occurring in Japan.

13.3.1 Hopes, Concerns, and Expectations

At the end of the study, I asked each participant about their hopes and concerns for their future. Perhaps unsurprisingly, eldercare (both their obligations to care for elders, and concerns about their own eldercare) featured prominently in their responses. Many participants saw themselves sandwiched between their mothers-in-law (whom many were caring for) and daughters-in-law (who likely would care for them). Perhaps based on their own experiences of caring for their in-laws, about 50% reported they did not want to impose such a burden on the next generation, although almost 80% still expected their daughters-in-law to care for them.

In Japan, great importance is placed on the *chōnan* (first-born son), in terms of carrying on the family name, ancestor worship, and business. When women marry, they are removed from their natal family register and added to their husband's register. In the case of families without sons, a daughter's husband will sometimes be added to her family register in place of the first-born son and will continue the male line. Pre-WWII, Japan had primogeniture with sole inheritance by the eldest son. This was abolished in the 1947 constitution, but the first-born son's wife is still expected to care for her parents-in-law in many families. Thus, while primogeniture no longer applies to money, it still seems to apply to obligations to care for parents. Increasingly, with decreasing birth rates, most sons are first-born sons, since families often have only one or two kids. Thus, non-first-born sons (i.e., those without parental care "strings attached") are increasingly hard to find.

Recently, intergenerational relationships have been changing. Whereas many women previously had issues and tensions with their mothers-in-law, increasingly mothers-in-law may be less heavy-handed. Instead, they're making efforts to help their daughters-in-law, reflecting shifts in women's positions within the household. That is, if women are lucky enough to have daughters-in-law. With decreasing marriage and birth rates and increasing numbers of unmarried adults in their 20s and 30s still living at home, concerns about their children's marriage prospects were prevalent. Several participants even asked me if I knew of any eligible potential partners for their children through my networks. I didn't feel very comfortable playing matchmaker, but I realized how desperate and concerned some of them must have felt to ask me.

In a culture with more strictly defined gender roles and expectations about male and female roles with respect to children and work outside the home, mothers' concerns about their children succeeding are understandable. Women's concerns about their children ranged from them not finding good jobs or partners to having children who never leave home. While women had concerns about their children's employment and marriage prospects, success in these areas often seemed to create new concerns, for example about gender roles.

13.3.2 Gender Roles and Marriage

Due to very long working hours and dedication to companies for those people (usually males) fortunate enough to get such jobs, wives were often expected to quit their jobs and careers, particularly when they had children, contributing to economic gender inequities and vulnerabilities particularly if divorce occurred.

Fieldnotes

I became quite close to one participant and her family, consisting of her husband and two daughters. Her daughters' stories highlight multiple factors influencing the roles available to women in Japan.

Her first-born daughter trained to be an English teacher. After teaching for several years, she married her college sweetheart and contented herself to being a full-time housewife for her salaryman husband. Large Japanese companies traditionally compensate their employees well enough so that their wives don't have to work outside the home. This allows them to support their husband's long working hours, particularly when they have children, and to follow them easily when they are transferred to new locations for work, a common practice in Japan's employment system. Once their third child entered school, she returned to teaching part-time.

Her second-born daughter trained as a nurse, but also quit once she got married, after an omiai *(arranged marriage meeting). It is still difficult for women to have children and continue working in their profession, especially if they are separated from extended family support. Government daycares are subsidized in Japan, but demand outpaces supply considerably in many areas, and many women do not want to work the long hours expected of them, particularly when their children are young. Many men might not want to work such long hours either, but they have less choice, although the rise of contract labor and the decrease of lifetime employment is changing this.*

Interestingly, decades before, the participant herself had continued working as a nurse while raising her two daughters. Importantly, while she and her daughters all married chōnan *(first-born sons), only she lived with her mother-in-law. And it was her mother-in-law who enabled her to continue working, helping to take care of the kids while she worked. Furthermore, her husband was a civil servant, and did not have to move for work, so she never had to quit her job to follow her husband.*

The factors influencing the lives of this participant and her daughters highlight the interplay of ultimate factors, including societal expectations and practices related to male vs. female labor, patterns of geographical proximity to extended family, professional roles available to women (e.g., teacher, nurse), and cultural expectations for married women. Culture plays a significant role in the timing of life histories, influencing fertility by cultural influences on age at marriage, which is increasing in Japan much to the chagrin of some of the older generation.

Fieldnotes

One participant's family business focused on traditional arts. Traditional values seemed important, and in talking about her husband, she often spoke of his samurai *(aristocratic warrior) values in a very respectful tone. Her husband was a man of great stature, both physical and social, with quite a presence. She often expressed concern about whether her children would have stable careers and find partners and have families.*

My participant's tatemae *(public façade) may have prevented her from discussing internal family struggles, but her admission, following the great Tohoku Earthquake in 2011, that concerns about the Fukushima nuclear plant were nothing compared to some of the family stress she had experienced over the years, was telling. She and her immediate family eventually moved out of the extended family compound and seemed content with all working together, with no outsiders having joined their family register and the daughters not having left theirs. They were an example of the changing nature of (and expectations about) families in 21st-century Japan.*

Despite stereotypes of gender and marital relations, diversity was apparent in my sample of women. Dynamics of relationships are hard to discern from the outside, and even among friends, people don't always know what transpires in each other's marriages.

Fieldnotes

In an unusual interview with two women friends (almost all other interviews were one-on-one), an interesting exchange ensued when I asked a question about frequency of marital fights. While one woman responded, "occasionally," the other woman loudly declared that she never fought with her husband. Her friend turned to her with a look of shock and surprise. And of course, my research assistant and I also very much wanted to know what her secret was. She paused, for effect, and said "We never talk. If we talked, we would fight," and then laughed heartily, as did we.

When marriages ended, different challenges might occur. I knew several Japanese women, some in their 20s and 30s, and several in their 50s or later, who got divorced, and their stories highlight the constraints and cultural expectations surrounding marriage and women's roles, even as some of these norms may be changing.

Fieldnotes

When she was in her 20s, one woman quit her full-time position when she got married. Her mother-in-law allowed her to continue working until she was able to get pregnant and have a baby, after which she was expected to quit to be a full-time mother. However, the relationship between the mother-in-law and her son contributed to problems in the marriage, and they eventually divorced. The cultural and political practice of women being placed on their husband's family register when they marry means that their children also belong to the father's registry alone. After her divorce, her mother-in-law informed her that she would only be allowed to see her infant son when he requested that!

In contrast, in a sign of changing ultimate cultural practices, when another woman in her 20s got divorced, she moved in with her own parents, changed her name back to her maiden name, and raised her son within her own natal family.

Not all living arrangements dissolved when marriages faltered. Japan also has had an arrangement called *kateinai rikon* (or within-house divorce) for at least several decades. More recently, people seem to be increasingly divorcing after *kōnenki*, referred to as *kōnenki rikon (kōnenki* divorce*)*. It is hard to know what relationships are like behind closed doors, and certainly hard to know what transpires in people's hearts and heads, i.e., what they really feel and think. Over the months and years, as I came to know many of these women, they increasingly opened up to me. I was privileged to hear not only some of their struggles, but also some of their hopes and dreams, and the journeys to make some of them come true.

The combination of long-life expectancy and changing gender and intergenerational roles of the 21st century appears to have created opportunities for many Japanese midlife women to reflect on the meaning of the next stage of their lives. When asked to compare their generation with that of their mothers, many women (>45%) described "freedom" and commented on changing gender relations and family dynamics.[16] As women move into the second half of their life during *kōnenki*, many seem to be embracing and exploring freedom in various forms. Several women mentioned *jibun rashiku ikiru*, which can be translated as living as your (full/true) self, or perhaps self-actualization. A *jibun rashī* (true self) life may entail caring less about societal expectations and instead following one's dreams.

Fieldnotes

The participant who was the first person to mention jibun rashī *(true self) life to me in the final interview, brought it up repeatedly in conversations in subsequent years. When I first met her, she was caring for elderly in-laws, while running a home-business, and fulfilling her duties as a wife. Years later, she had suffered some health problems and decided to actively seek a* jibun rashī *life, which for her involved doing art and spiritual exploration. Every time I met her, she seemed to have found greater inner peace, and exuded contentment and wisdom, all while continuing her support of and involvement with her family, albeit seemingly with a new, more independent dynamic vis-à-vis her husband.*

Many of the women who during or after *kōnenki* seemed to embrace a search for a *jibun rashī* life were people whose identities for much of their lives derived from serving their families. And while many continued to serve others, their choices seemed influenced more by their own authentic paths.

Fieldnotes

When I met her, one participant was raising three children, caring for not only her in-laws but also her own parents, and volunteering regularly. She was probably one of the most organized, disciplined people

I have ever met, doing everything to an incredibly high standard. Once her children were grown and left home, she went back to school to train in the health professions. She moved from her traditional home to a new modern home. When I visited her at her new home, I saw her in a new light (literally and figuratively), energized by her new passions and exploration, applying her curious mind to the study of a healing art.

The phrase *jibun rashiku ikiru* is also related to the Okinawan concept of *ikigai* (life's meaning or purpose), a hallmark of many people living in the Blue Zones. Many of the women I interviewed demonstrate that just because you reach the end of reproduction or maybe even the end of your paid work/career doesn't mean you have to give up on *ikigai*. Perhaps for many it is the opposite. A lot of people post-retirement do find, perhaps for the first time, what their *ikigai* is. Tamashio's excellent Ted Talk[17] about *ikigai* brings this discussion full circle back to the Blue Zones of Chapter 12, as it connects the concept to healthy aging. *Ikigai* may provide motivation for healthy eating and social interaction, which reinforce each other, with positive feedback loops.

The ability to envision and explore one's *ikigai* (life's meaning and purpose) in order to *jibun rashiku ikiru* (to live for/as oneself) depends on larger societal factors, having key social determinants of health and basic needs met, and role expectations for women and older people that allow and support flexibility and exploration. While *ikigai* may be a secret to healthy aging and longevity, it is also useful for readers of any age to consider no matter what chapter of their life they are in. Creating ultimate level societal factors that support the "self-actualization" of oneself and our communities requires an understanding of how political and social economic determinants, expectations for changing gender and age roles, and cultural framings (e.g., medicalization) of signs and symptoms of aging all intersect to affect health and quality of life throughout the life course.

References

1. Drescher J. 2015. Out of DSM: Depathologizing homosexuality. *Behav Sci (Basel)*. 5(4):565–575. https://doi.org/10.3390/bs5040565
2. Melby MK, Lock M, Kaufert P. 2005. Culture and symptom reporting at menopause. *Hum Reprod Update*. 11(5):495–512. https://doi.org/10.1093/humupd/dmi018
3. Avis NE, Ory M et al. 2003. Health-related quality of life in a multiethnic sample of middle aged women: Study of Women's Health Across the Nation (SWAN). *Med Care*. 41(11):1262–1276. https://doi.org/10.1097/01.MLR.0000093479.39115.AF

4. Dennerstein L, Lehert P et al. 2007. A symptomatic approach to understanding women's health experiences: A cross-cultural comparison of women aged 20 to 70 years. *Menopause.* 14(4):688–696. https://doi.org/10.1097/gme.0b013e31802dabf0
5. Lock M. 1993. *Encounters with Aging: Mythologies of menopause in Japan and North America.* Berkeley, CA: University of California Press.
6. Foxcroft L. 2009. *Hot Flushes, Cold Science.* London: Granta Books.
7. Jermyn D. 2024. Whose menopause revolution? Investigating the UK's 'Davina effect' and the contemporary menopause market. *Eur J Cult Stud.* 0(0): 13675494241287931. https://doi.org/10.1177/13675494241287931
8. Martin E. 1991. The egg and the sperm: How science has constructed a romance based on stereotypical male-female roles. *Signs.* 16(3):485–501. https://www.jstor.org/stable/3174586
9. Miller G (host). 2009, March 17). Menopause and medicine. Louise Foxcroft: Hot Flushes, Cold Science. (No 25) [Audio podcast episode]. In Podularity Podcast. https://podularity.com/2009/03/17/25-menopause-and-medicine/
10. Lock M. 1980. *East Asian Medicine in Urban Japan.* Berkeley, CA: University of California Press.
11. Yasui T, Ideno Y et al. 2022. Prevalence of the use of oral contraceptives and hormone replacement therapy in Japan: The Japan nurses' health study. *J Epidemiol.* 32(3):117–124. https://doi.org/10.2188/jea.JE20200207
12. Rosenberger NR. 1992. The process of discourse: Usages of a Japanese medical term. *Soc Sci Med.* 34(3):237–247. https://doi.org/10.1016/0277-9536(92)90266-S
13. Harlow SD, Crawford SL, et al. 2000. Self-defined menopausal status in a multi-ethnic sample of midlife women. *Maturitas.* 36(2):93–112. https://doi.org/10.1016/s0378-5122(00)00145-6
14. Busch H, Barth-Olofsson AS et al. 2003. Menopausal transition and psychological development. *Menopause.* 10(2):179–187. https://doi.org/10.1097/00042192-200310020-00011
15. Hvas L. 2006. Menopausal women's positive experience of growing older. *Maturitas.* 54(3):245–251. https://doi.org/10.1016/j.maturitas.2005.11.006
16. Melby MK. 2016. Lessons on aging: Hopes and concerns of Japanese women at midlife. *J Women Aging.* 28(2):127–140. https://doi.org/10.1080/08952841.2014.950544
17. Tamashio T. 2018. How to Ikigai. *TEDxYYC.* Available at: https://www.youtube.com/watch?v=pk-PcJS2QaU. Accessed 9 Feb, 2025.

14

QUESTIONS AND FUTURE EXPLORATIONS

The proximate-ultimate biocultural framework provides a structure for thinking about complementary levels of analysis. Multiple levels of explanation are needed because no single level tells the whole story. Insights from complementary levels all provide important pieces of the puzzle, whether that puzzle is aging or menopause, or other health phenomena like depression or diabetes.

With health issues, we often focus on proximate biology because we are typically trying to solve an immediate problem, rather than understand its ultimate causes. This easily reinforces the universalism and biological determinism of biomedicine. But few health conditions are solely biological and most are influenced by culture to create local biologies. Understanding proximate mechanisms and how things work is important, but humans often want to know why. Why do they get sick? Why now? Why do some people but not others have symptoms at certain times in the life course? These questions are important not just to satisfy our curiosity, but to potentially give us better ways to solve problems, for example of mismatch between our evolutionary heritage and our contemporary lifestyles, and ideally prevent them.

Just as the Blue Zones (note the plural) demonstrate that there is not a "one size fits all" diet or lifestyle for healthy aging, there is not a "one size fits all" way to navigate the menopausal transition or postmenopausal period. Individual variation as well as population- and cultural-level variation are important to consider, and these are often heavily influenced by preceding life history and interactions with biocultural environments over the life course.

DOI: 10.4324/9781003644521-18

I often get asked "How does medical anthropology differ from medicine?" Medicine aims to cure people of diseases, or solve clinical problems, whereas medical anthropology tries to understand them. Medical anthropology also acknowledges, and delights in, the variation in human ways of living and interacting with their environment, and celebrates how eclectic the field is in drawing from so many disciplines with the aim of achieving a holistic understanding. By expanding our perspective on aging and menopause beyond the proximate biology quadrant (typically the focus of biomedicine), to include factors in the other quadrants we can consider biological and cultural variation and the many paths to local biologies and experiences, as opposed to a single universal global biology and experience of aging and menopause (Figure 14.1).

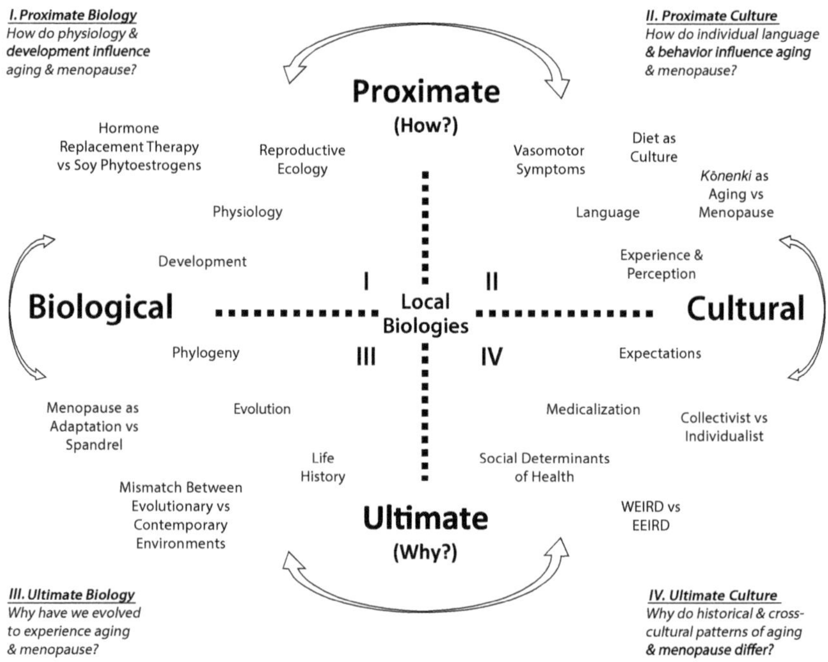

FIGURE 14.1 Graphical summary highlighting key themes and issues from each quadrant (labeled in center with Roman numerals) of the proximate-ultimate biocultural framework on aging and menopause examined in this book. The concept of local biologies is at the center, influenced by factors from all quadrants. Three key themes discussed in each quadrant (closer to center) are listed, with key tensions or questions highlighted around the outside. Primary questions from each quadrant introduced at the opening of each part of the book are shown in the corners.

14.1 Summary of Quadrants

In Part I, as we explored the proximate biology quadrant, we discussed understandings, heavily influenced by biomedicine, of the "normal" physiology of aging and menopause. The presentation in Part I of the "standard" or "normal" view allows us to challenge and complicate the view in subsequent parts of the book, as we integrated more ultimate and cultural perspectives. Biology is not destiny, and many biocultural factors influence proximate biological manifestations. For aging, several mechanisms may operate simultaneously on different levels of biology ranging from genes to hormones and free radicals. The calcium and osteoporosis paradoxes challenge the idea of universal biology and highlight the need for consideration of cultural and environmental factors as well as ultimate perspectives. For menopause, expanding our perspective from a narrow biomedical perspective that looks primarily at hormones to a perspective considering developmental factors and variation in reproductive ecology (e.g., number and timing of pregnancies and duration of breastfeeding, and exposure to hormone therapy in the form of birth control) as well as the autonomic nervous system may broaden our understanding of factors influencing later life experience of menopause and treatments beyond hormone therapy.

In Part II, the proximate culture quadrant, we examined variation in culture that juxtaposes the insider/*emic* concept of *kōnenki* with outsider/*etic* concepts of aging and menopause. We examined culture in the form of diet with a focus on soy that may influence local biologies. The language of hot flushes may also provide clues to nuanced experiences and perceptions of local biologies and raises questions about whether experiences or perceptions of hot flushes differ cross-culturally. Do Japanese hot flush terms hint at differing underlying physiology? Or do Japanese hot flush terms suggest different perceptions and categorizations of a continuum of thermoregulatory experience that Western, Educated, Industrialized, Rich and Democratic (WEIRD) cultures see in more binary terms (i.e., hot flush or not)? The symptom of chilliness expands our perspective on the thermoregulatory continuum, possibly influenced by diet, which is of course both cultural and biological, and influenced by evolutionary heritage, geography, and history.

In Part III, the ultimate biology quadrant, we examined evolutionary theories for aging and menopause, with a focus on life history, and alternative explanations to adaptation such as phylogenetic constraints and mismatch. Later in the lifespan, selection is weak, and thus it is important to consider tradeoffs with priorities (of maximizing fitness) earlier in the lifespan. Genes are important, but whose genes: Host or microbial genes? Microbes may allow humans to adapt on faster time scales. Genes may exert positive effects earlier in the life course but negative effects later in life (antagonistic pleiotropy). Such tradeoffs at the level of genes, combined with developmental constraints and mismatches between our evolutionary and contemporary environments, suggest that aging- and menopause-related symptoms may be best understood and approached not only with a life course perspective but a longer evolutionary life history perspective. Menopause may be more accurately thought of as an evolutionary spandrel, revealed in females by the separation of the following two "arches": (1) an adaptive extension of the arch of somatic lifespan, which in males is tightly linked to reproductive lifespan, since males can directly increase their fitness by reproducing at later ages; and (2) an arch of female reproductive function, which is under phylogenetic and developmental constraints to maximize earlier reproduction (an old mammalian pattern). This has implications for how we treat aging and menopause, culturally and biologically.

In Part IV, the ultimate culture quadrant, we examined WEIRD vs. Eastern, Educated, Industrialized, Rich, and Democratic (EEIRD) cultures, a comparison characterized by differences between individualism and collectivism. These differences have implications for the importance of the role played by social connection in the attainment of healthy aging and navigation of menopause. Similarly, social determinants of health play a role in Japan's long-life expectancy, although other populations and locations (Blue Zones) demonstrate that there is not a single path to healthy aging. However, there are broad patterns shared by populations that achieve healthy aging. Medicalization influences expectations about aging and menopause, which in turn influence perceptions and experience of symptoms. Comparative analyses show us factors we may have ignored from the vantage point of our own culture and demonstrate the range of human possibilities of the experience of aging and menopause.

14.2 Frameworks in Comparison

The proximate-ultimate biocultural framework also yields insights when used to reflect on other frameworks, some of which I explore below (Roman numerals refer to the quadrants of the proximate-ultimate biocultural framework). For example, the Disease (I)-Illness (II)-Sickness (IV) framework (Chapter 2) covers three of the four quadrants but does not usually include evolutionary (III) considerations. Disease refers to proximate biological phenomena, usually the observable clinical manifestations of (assumed) abnormality of physical function, categorized by a health professional. Illness describes the proximate culture of subjective experience, perceptions, and lived experience (including symptoms). Sickness encompasses ultimate cultural and social phenomena, roles, and factors associated with illness.

The river parable commonly used in public health (Chapter 11) discusses upstream (ultimate) social determinants of health (IV) and their downstream proximate biological consequences and manifestations (I) which can be mediated by more proximate cultural factors (II). But the river parable ignores evolutionary (III) considerations. Evolutionary medicine attempts to integrate evolutionary perspectives (III), such as adaptation, constraints, and mismatches with more proximate biological (I) concerns of physiological and anatomical manifestations of disease.

Lock's concepts of local biology and "situated biologies" explicitly examine how social and historical/political/economic context (IV) might produce "local" biologies (see center of Figure 14.1). While she did not collect biological data in her research, she suggested that soy phytoestrogens might play a role in Japan. And later, she explored epigenetics and the microbiome, factors that influence how cultural factors (II and IV) become embodied and lead to changes in physiology and health (I). The proximate-ultimate biocultural framework described in this book builds on these other frameworks, and expands them to include and integrate four quadrants along proximate-ultimate and biocultural axes, leading to more holistic approaches and perspectives.

14.3 Blurring of Lines – Interactions between and across Quadrants

Just as cross-cultural analyses often lead to interesting questions and important insights, examining cross-quadrant interactions and connections may lead to important questions. While the framework and quadrants provide a structure to understand complementary levels of analysis, they are a simplification of reality. There are phenomena that don't fit neatly into any quadrant, and that is wonderful, because it shows that these levels and quadrants are all interconnected and needed for holistic understanding. Figure 14.1 shows the framework with axes using dotted lines, indicating that they are porous, conveying potential interaction, overlap, and influence.

Proximate approaches answer questions of how things work, while ultimate approaches answer questions of why things are the way they are. Biological and cultural levels of analysis are often viewed as very different domains. But lines between quadrants can and do blur, and the interconnections generate interesting and important questions, several of which are described below (Roman numerals refer to the quadrants of the proximate-ultimate biocultural framework). Readers are encouraged to think of others.

- *(I–II)* Proximate Biology – Proximate Culture: Is *kōnenki* more biological or cultural? Is *kōnenki* more of a phenomenon of aging or menopause? How can one word or concept mean such different things, and what are the implications for how people navigate this experience? What are the treatment implications? Is it unique to women or can men experience it, and if so, does that mean it is more cultural or can it be biological as well? What are the implications for biomedical understandings of menopause?

- (III–I) Ultimate Biology – Proximate Biology: How do phylogenetic constraints and life history influence the physiology of aging and menopause? What are the implications for the design of proximate biological interventions (e.g., hormonal contraception), as well as biocultural factors such as diet and activity levels?
- (I–IV) Proximate Biology – Ultimate Culture: How do laboratory insights from proximate biological mechanisms of caloric restriction in primates relate to the Blue Zone Okinawan *hara hachi-bu* (eat until you are 80% full) lifestyle guidance and approach? What are the implications for healthy aging policy and clinical interventions?
- (IV–II) Ultimate Culture – Proximate Culture: How might ultimate cultural factors of medicalization, expectations about aging, and collectivist vs. individualist orientations constrain and moderate proximate culture and individual behaviors like diet and health behaviors, and influence which symptoms are noticed or perceived as bothersome?
- (IV–III) Ultimate Culture – Ultimate Biology: How do ultimate culture perspectives of (primarily WEIRD) scientists influence which evolutionary hypotheses about aging and menopause are considered? Could historical and cultural views of women primarily as reproducers influence scientists' focus on finding an adaptive evolutionary explanation for menopause, rather than constraints or mismatch with current environments? Why might we prefer explanations that reinforce gender roles?
- (II–I) Proximate Culture – Proximate Biology: How might language, for example the multiple terms for hot flushes in Japan, provide clues to underlying proximate biological phenomena that do not exist in WEIRD cultures, or perhaps simply lack labels in biomedicine?
- (IV–II) Ultimate Culture – Proximate Culture: How do expectations about *kōnenki* as a period of hormone imbalance influence the experience in contrast to cultures that view menopause as a loss of hormones? Do expectations associated with words such as "imbalance" vs. "loss" affect the total experience?
- (IV–III) Ultimate Culture – Ultimate Biology How might WEIRD cultural factors (e.g., parental leave policies) influence life history, such as timing and frequency of reproduction and breastfeeding?
- (III/IV–I) Ultimate Biology/Culture– Proximate Biology: How do ultimate biocultural factors such as climate, built environment, development, physiology, previous exposure to external hormones, and reproductive history influence which proximate biology symptoms manifest when we get older?
- (III–IV–I–II) Ultimate Biology – Ultimate Culture – Proximate Biology – Proximate Culture: How do ultimate cultural factors of modernization lead to a mismatch with our evolutionary history? In turn, how do developmental constraints on reproductive functioning and life history tradeoffs in novel evolutionary contexts influence our proximate biology symptoms and proximate culture experience of aging and menopause?

Questions about how culture influences health are important to ask, but it is also important to ask how the culture of those asking the questions plays a role. To what extent can scientists' individual experiences and backgrounds influence the questions they ask, and how and why they ask them, and how the answers are interpreted? Our scientific knowledge is influenced, and constrained, by the training and interests of scientists, who are products of their culture. While we can study other cultures and expand our perspectives, involving people from multiple cultures and disciplinary backgrounds in research can also serve to expand the questions that are asked and the approaches that are used.

All questions are posed by humans who are influenced by their culture – in terms of what they see and don't see – their interests, hypotheses, biases, and blind spots. Recall my story of forgetting to ask sex or gender on the Kanazawa symptom checklist, because I was adapting a survey for menopausal women, even though I explicitly wanted to include men's perspectives! Perhaps if I had included a man on my research team, or at least piloted the questionnaire with a man, he might have realized my mistake and saved me from a lot of extra work.

Might scientists who grow up and work in WEIRD and more individualistic cultures be more prone to looking for individual causes of variation in experience of aging and menopause, as opposed to more social, collectivist influences? My hope is that readers will find this book's framework useful and be inspired to consider more levels of analysis and more perspectives when considering questions that interest them. I encourage readers to think about other ways in which the framework's lines blur and are really porous, and to consider interdisciplinary, proximate-ultimate and biocultural interconnections. In doing so, hopefully you will be inspired to find your own *ikigai* (life's purpose), and expand your ways of thinking about, inquiring about, experiencing, and acting in the world.

BIBLIOGRAPHY

Adlercreutz, H., and W. Mazur. 1997. "Phyto-oestrogens and western diseases." *Ann Med.* 29(2):95–120.

ACS. 2025. "Key statistics for melanoma skin cancer." *American Cancer Society*. Accessed 21 May 2025. https://www.cancer.org/cancer/types/melanoma-skin-cancer/about/key-statistics.html.

AFAR. 2016. *Theories of aging. Accessed* 21 May 2025. https://www.afar.org/imported/AFAR_INFOAGING_GUIDE_THEORIES_OF_AGING_2016.pdf.

Alder, E. 1998. "The Blatt-Kupperman menopausal index: a critique." *Maturitas* 29 (1): 19–24. https://doi.org/10.1016/S0378-5122(98)00024-3.

Applegate, C. C., J. L. Rowles, K. M. Ranard, S. Jeon, and J. W. Erdman. 2018. "Soy consumption and the risk of prostate cancer: an updated systematic review and meta-analysis." *Nutrients* 10 (1):1–25. https://doi.org/10.3390/nu10010040.

Avis, N. E., M. Ory, K. A. Matthews, M. Schocken, J. Bromberger, and A. Colvin. 2003. "Health-related quality of life in a multiethnic sample of middle-aged women: Study of Women's Health across the Nation (SWAN)." *Med Care* 41 (11): 1262–1276. https://doi.org/10.1097/01.MLR.0000093479.39115.AF.

Avis, N. E., R. Stellato, S. Crawford, J. Bromberger, P. Ganz, V. Cain, and M. Kagawa-Singer. 2001. "Is there a menopausal syndrome? Menopausal status and symptoms across racial/ethnic groups." *Soc Sci Med* 52 (3): 345–56. https://doi.org/10.1016/s0277-9536(00)00147-7.

Bapteste, E., P. Huneman, L. Keller, J. Teulière, P. Lopez, E. C. Teeling, A. B. Lindner, A. Baudisch, W. B. Ludington, and C. Franceschi. 2023. "Expanding evolutionary theories of ageing to better account for symbioses and interactions throughout the Web of Life." *Ageing Res Rev* 89: 101982. https://doi.org/10.1016/j.arr.2023.101982.

BBC. 2021. "Climate change: do I need to stop eating meat?". *BBC*. Accessed 21 May 2025. https://www.bbc.co.uk/news/explainers-59232599.

Beetch, M., S. Harandi-Zadeh, K. Shen, K. Lubecka, D. D. Kitts, H. M. O'Hagan, and B. Stefanska. 2020. "Dietary antioxidants remodel DNA methylation patterns in chronic disease." *Br J Pharmacol* 177 (6): 1382–1408. https://doi.org/10.1111/bph.14888. https://bpspubs.onlinelibrary.wiley.com/doi/abs/10.1111/bph.14888.

Bergeron, L. A., S. Besenbacher, J. Zheng, P. Li, M. F. Bertelsen, B. Quintard, J. I. Hoffman, Z. Li, J. St Leger, C. Shao, J. Stiller, M. T. P. Gilbert, M. H. Schierup, and G. Zhang. 2023. "Evolution of the germline mutation rate across vertebrates." *Nature* 615 (7951): 285–291. https://doi.org/10.1038/s41586-023-05752-y.

Best, E. V., and M. D. Schwartz. 2014. "Fever." *Evo Med Public Health* 2014 (1): 92–92. https://doi.org/10.1093/emph/eou014.

Beyene, Y. 1986. "Cultural significance and physiological manifestations of menopause: A biocultural analysis." *Culture, Medicine and Psychiatry*. 10:47–71.

Bhanote, M. 2022. "Western vs. Eastern medicine: what to know." *Medical News Today*. Accessed 21 May 2025. https://www.medicalnewstoday.com/articles/western-vs-eastern-medicine.

Bivens, J., and J. Kandra. 2022. "CEO pay has skyrocketed 1460% since 1978." *Economic Policy Institute*. Accessed 21 May 2025. https://www.epi.org/publication/ceo-pay-in-2021/

Bogin, B. 2015. "Ch 20. Human Growth and Development." In *Basics in Human Evolution, Meuhlenbein*, MP (ed). 285–293. London: Academic Press.

Bondesson, M., and J.-A. Gustafsson. 2010. "Does consuming isoflavones reduce or increase breast cancer risk?" *Genome Med* 2 (12): 90. https://doi.org/10.1186/gm211.

Brown, N. M., and K. D. R. Setchell. 2001. "Animal models impacted by phytoestrogens in commercial chow: implications for pathways influenced by hormones." *Lab Invest* 81 (5): 735–747. https://doi.org/10.1038/labinvest.3780282.

Buettner, D. 2013. "How to live to be 100+." Ted-Ed. Accessed 21 May 2025. https://www.ted.com/talks/dan_buettner_how_to_live_to_be_100?language=en

Buettner, D., and S. Skemp. 2016. "Blue zones: lessons from the world's longest lived." *Am J Lifestyle Med* 10 (5): 318–321. https://doi.org/10.1177/1559827616637066.

Busch, H., A. S. Barth-Olofsson, S. Rosenhagen, and A. Collins. 2003. "Menopausal transition and psychological development." *Menopause* 10 (2): 179–187. https://journals.lww.com/menopausejournal/abstract/2003/10020/menopausal transition_and_psychological.11.aspx.

Butwick, A. J., J. Bentley, C. A. Wong, J. M. Snowden, E. Sun, and N. Guo. 2018. "United States state-level variation in the use of neuraxial analgesia during labor for pregnant women." *JAMA Netw Open* 1 (8): e186567. https://doi.org/10.1001/jamanetworkopen.2018.6567.

Byars, S. G., Q. Q. Huang, L. A. Gray, A. Bakshi, S. Ripatti, G. Abraham, S. C. Stearns, and M. Inouye. 2017. "Genetic loci associated with coronary artery disease harbor evidence of selection and antagonistic pleiotropy." *PLoS Genet* 13 (6): e1006328. https://doi.org/10.1371/journal.pgen.1006328.

Cagnacci, A., and M. Venier. 2019. "The controversial history of hormone replacement therapy." *Medicina (Kaunas)* 55 (9). https://doi.org/10.3390/medicina55090602.

Campbell, K. L., and J. W. Wood. 1994. "Introduction: what is human reproductive ecology, and why should we care about studying it?" *Ann NY Acad Sci* 709 (1): 1–8. https://doi.org/10.1111/j.1749-6632.1994.tb30384.x .

Carr, B. R., and J. D. Wilson. 1987. "Disorders of the ovary and female reproductive tract." In *Harrison's principles of internal medicine*, edited by E Braunwald KJ, Isselbacher, RG Petersdorf et al, 1818–1837. New York: McGraw-Hill.

Chevalley, T., and R. Rizzoli. 2022. "Acquisition of peak bone mass." *Best Pract Res Clin Endocrinol Metab* 36 (2): 101616. https://doi.org/10.1016/j.beem.2022.101616.

Chiao, J. Y., T. Harada, H. Komeda, Z. Li, Y. Mano, D. Saito, T. B. Parrish, N. Sadato, and T. Iidaka. 2009. "Neural basis of individualistic and collectivistic views of self." *Hum Brain Mapp* 30 (9): 2813–2820. https://doi.org/10.1002/hbm.20707.

Choi, S.-W., and S. Friso. 2010. "Epigenetics: a new bridge between nutrition and health." *Adv Nutr* 1 (1): 8–16. https://doi.org/10.3945/an.110.1004.

Cohen, A. A., C. F. D. Coste, X.-Y. Li, S. Bourg, and S. Pavard. 2020. "Are trade-offs really the key drivers of ageing and life span?" *Funct Ecol* 34 (1): 153–166. https://doi.org/10.1111/1365-2435.13444

Cohen, A. A., L. Ferrucci, T. Fülöp, D. Gravel, N. Hao, A. Kriete, M. E. Levine, L. A. Lipsitz, M. G. M. Olde Rikkert, A. Rutenberg, N. Stroustrup, and R. Varadhan. 2022. "A complex systems approach to aging biology." *Nat Aging* 2 (7): 580–591. https://doi.org/10.1038/s43587-022-00252-6.

CPC. 2011. "Pyramids App: the dynamics of populations structures." Accessed 21 May 2025. https://pyramids.cpc.ac.uk/.

Craft, L. 2023. "Japan's government will play romantic matchmaker in a bid to shore up the birth rate. Experts say it will fail." *CBS News*. Accessed 4 Feb 2025. https://www.cbsnews.com/news/japan-birth-rate-population-decline-government-matchmaking-marriage-children/

Croft, D. P., L. J. N. Brent, D. W. Franks, and M. A. Cant. 2015. "The evolution of prolonged life after reproduction." *Trends Ecol Evol* 30 (7): 407–416. https://doi.org/10.1016/j.tree.2015.04.011.

Dańko, M. J., J. Kozłowski, and R. Schaible. 2015. "Unraveling the non senescence phenomenon in Hydra." *J Theoret Biol* 382: 137–149. https://doi.org/10.1016/j.jtbi.2015.06.043. https://www.sciencedirect.com/science/article/pii/S0022519315003227.

Darwin, C. 1871. *The descent of man, and selection in relation to sex*. London: J. Murray.

Dato, S., G. Rose, P. Crocco, D. Monti, P. Garagnani, C. Franceschi, and G. Passarino. 2017. "The genetics of human longevity: an intricacy of genes, environment, culture and microbiome." *Mech Ageing Dev* 165 (Pt B): 147–155. https://doi.org/10.1016/j.mad.2017.03.011.

Dattani, S. 2023. "How does the risk of death change as we age – and how has this changed over time?". *Our World in Data*. Accessed 22 May 2025. https://ourworldindata.org/how-do-the-risks-of-death-change-as-people-age.

Dattani, S., L. Rodés-Guirao, H. Ritchie, E. Ortiz-Ospina, and M. Roser. 2023. "Life expectancy." Accessed 23 Jan. https://ourworldindata.org/life-expectancy.

Davenport, L. 2020. "The fluidity of racial classifications." *Annu Rev Political Sci* 23: 221–240 https://doi.org/10.1146/annurev-polisci-060418-042801.

de Kat, A. C., F. Roelofs, P. E. Slagboom, F. J. M. Broekmans, M. Beekman, and N. van den Berg. 2024. "Late reproduction is associated with extended female survival but not with familial longevity." *Reprod BioMed Online* 49 (3): 104073. https://doi.org/10.1016/j.rbmo.2024.104073.

Deecher, D. C., and K. Dorries. 2007. "Understanding the pathophysiology of vasomotor symptoms (hot flushes and night sweats) that occur in perimenopause, menopause, and postmenopause life stages." *Arch Womens Ment Health* 10 (6): 247–257. https://doi.org/10.1007/s00737-007-0209-5.

Dennerstein, L., P. Lehert, P. E. Koochaki, A. Graziottin, S. Leiblum, and J. L. Alexander. 2007. "A symptomatic approach to understanding women's health experiences: a cross-cultural comparison of women aged 20 to 70 years." *Menopause* 14 (4): 688–696. https://doi.org/10.1097/gme.0b013e31802dabf0.

Dettwyler, K. A. 1995. "A time to wean: the hominid blueprint for the natural age of weaning in modern human populations." In *Breastfeeding: biocultural perspectives*, Stuart-Macadam, Patricia Dettwyler and Katherine A (eds), 39–74. New York: Aldine De Gruyter.

Diogo, R. 2018. "First detailed anatomical study of bonobos reveals intra-specific variations and exposes just-so stories of human evolution, bipedalism, and tool use." *Front Ecol Evol* 6. https://doi.org/10.3389/fevo.2018.00053.

Doi, T. 1973. *The anatomy of dependence*. Translated by John Bester. Tokyo: Kodansha International.

Drescher, J. 2015. "Out of DSM: depathologizing homosexuality." *Behav Sci (Basel)* 5 (4): 565–575. https://doi.org/10.3390/bs5040565.

Eaton, S. B., M. Konner, and M. Shostak. 1988. "Stone agers in the fast lane: chronic degenerative diseases in evolutionary perspective." *Am J Med* 84 (4): 739–749. https://doi.org/10.1016/0002-9343(88)90113-1.

Eaton, S. B., M. C. Pike, R. V. Short, N. C. Lee, J. Trussel, R. A. Hatcher, J. W. Wood, C. M. Worthman, N. G. Blurton Jones, M. J. Konner, K. R. Hill, R. Bailey, and A. M. Hurtado. 1994. "Women's reproductive cancers in evolutionary context." *Q Rev Biol* 69 (3): 353–367. https://doi.org/10.1086/418650.

Efferth, T., M. Banerjee, and N. W. Paul. 2017. "Broken heart, tako-tsubo or stress cardiomyopathy? Metaphors, meanings and their medical impact." *Int J Cardiol* 230: 262–268. https://doi.org/10.1016/j.ijcard.2016.12.129.

Ellis, S., D. W. Franks, M. L. K. Nielsen, M. N. Weiss, and D. P. Croft. 2024. "The evolution of menopause in toothed whales." *Nature* 627 (8004): 579–585. https://doi.org/10.1038/s41586-024-07159-9.

Ellison, P. T. 1994. "Advances in human reproductive ecology." *Annu Rev Anthropol* 23: 255–275. https://doi.org/10.1146/annurev.an.23.100194.001351.

Ellison, P. T., C. Panter-Brick, S. F. Lipson, and M. T. O'Rourke. 1993. "The ecological context of human ovarian function." *Hum Reprod* 8 (12): 2248–2258. https://doi.org/10.1093/oxfordjournals.humrep.a138015.

Etkin, N. L. 2006. *Edible medicines: an ethnopharmacology of food*. Tucson: University of Arizona Press.

Evolution, UC Museum of Paleontology Understanding. "Evolution 101." *Understanding Evolution*. Accessed 9 Feb 25. https://evolution.berkeley.edu/evolution-101/.

Ferrucci, L., F. Giallauria, and J. M. Guralnik. 2008. "Epidemiology of aging." *Radiol Clin North Am* 46 (4): 643–652. https://doi.org/10.1016/j.rcl.2008.07.005.

Finch, C. E. 2007. *The biology of human longevity: inflammation, nutrition and aging in the evolution of lifespans*. Burlington, MA: Academic Press.

Finch, C., and R. Rose. 1995. "Hormones and the physiological architecture of life history evolution." *Quart Rev Biol* 70: 1–52. https://doi.org/10.1086/418864.

Finney, C. A., N. W. Proschogo, A. Snepvangers, N. M. Holmes, R. F. Westbrook, and K. J. Clemens. 2020. "The effect of standard laboratory diets on estrogen signaling and spatial memory in male and female rats." *Physiol Behav* 215: 112787. https://doi.org/10.1016/j.physbeh.2019.112787.

Flanagan, E. W., J. Most, J. T. Mey, and L. M. Redman. 2020. "Calorie restriction and aging in humans." *Annu Rev Nutr* 40: 105–133. https://doi.org/10.1146/annurev-nutr-122319-034601.

Flatt, T., and L. Partridge. 2018. "Horizons in the evolution of aging." *BMC Biol* 16 (1): 93. https://doi.org/10.1186/s12915-018-0562-z.

Foxcroft, L. 2009. *Hot flushes, cold science*. London: Granta Books.

Frassetto, L. A., K. M. Todd, R. C. Morris, Jr., and A. Sebastian. 2000. "Worldwide incidence of hip fracture in elderly women: relation to consumption of animal and vegetable foods." *J Gerontol A Biol Sci Med Sci* 55 (10): M585–M592. https://doi.org/10.1093/gerona/55.10.m585.

Freedman, R. R. 2001. "Physiology of hot flashes." *Am J Hum Biol* 13 (4): 453–464. https://doi.org/10.1002/ajhb.1077.

Freedman, R. R. 2005. "Hot flashes: behavioral treatments, mechanisms, and relation to sleep." *Am J Med* 118 (12 Suppl 2): 124–130. https://doi.org/10.1016/j.amjmed.2005.09.046.

Freedman, R. R. 2014. "Menopausal hot flashes: mechanisms, endocrinology, treatment." *J Steroid Biochem Mol Biol* 142: 115–120. https://doi.org/10.1016/j.jsbmb.2013.08.010.

Freeman, E. W., K. E. Ensrud, J. C. Larson, K. A. Guthrie, J. S. Carpenter, H. Joffe, K. M. Newton, B. Sternfeld, and A. Z. LaCroix. 2015. "Placebo improvement in pharmacologic treatment of menopausal hot flashes: time course, duration, and predictors." *Psychosom Med* 77 (2): 167–175. https://doi.org/10.1097/psy.0000000000000143.

French, E. B., J. McCauley, M. Aragon, P. Bakx, M. Chalkley, S. H. Chen, B. J. Christensen, H. Chuang, A. Côté-Sergent, M. De Nardi, E. Fan, D. Échevin, P.-Y. Geoffard, C. Gastaldi-Ménager, M. Gørtz, Y. Ibuka, J. B. Jones, M. Kallestrup-Lamb, M. Karlsson, T. J. Klein, G. de Lagasnerie, P.-C. Michaud, O. O'Donnell, N. Rice, J. S. Skinner, E. van Doorslaer, N. R. Ziebarth, and E. Kelly. 2017. "End-of-life medical spending in last twelve months of life is lower than previously reported." *Health Aff* 36 (7): 1211–1217. https://doi.org/10.1377/hlthaff.2017.0174.

Fuentes, N., and P. Silveyra. 2019. "Estrogen receptor signaling mechanisms." *Adv Protein Chem Struct Biol* 116: 135–170. https://doi.org/10.1016/bs.apcsb.2019.01.001.

Gabriel, A. S., K. Ninomiya, and H. Uneyama. 2018. "The role of the Japanese traditional diet in healthy and sustainable dietary patterns around the world." *Nutrients* 10 (2). https://doi.org/10.3390/nu10020173.

Gapminder. 2025. "Gapminder resources." Accessed 1 Feb 2025. https://www.gapminder.org/resources/.

Ghosh, T. S., F. Shanahan, and P. W. O'Toole. 2022. "The gut microbiome as a modulator of healthy ageing." *Nat Rev Gastroenterol Hepatol* 19 (9): 565–584. https://doi.org/10.1038/s41575-022-00605-x.

Gibson, M. 2015. "The long, strange history of birth control." *Time*. https://time.com/3692001/birth-control-history-djerassi/.

Gilbert, J. A., M. J. Blaser, J. G. Caporaso, J. K. Jansson, S. V. Lynch, and R. Knight. 2018. "Current understanding of the human microbiome." *Nat Med* 24 (4): 392–400. https://doi.org/10.1038/nm.4517.

Gillies, G. E., and S. McArthur. 2010. "Estrogen actions in the brain and the basis for differential action in men and women: a case for sex-specific medicines." *Pharmacol Rev* 62 (2): 155–198. https://doi.org/10.1124/pr.109.002071.

"Global Gender Gap Report 2023." 2023. *World Economic Forum*. Accessed 4 Feb. https://www.weforum.org/publications/global-gender-gap-report-2023/in-full/benchmarking-gender-gaps-2023/

Gold, E. B. 2011. "The timing of the age at which natural menopause occurs." *Obstet Gynecol Clin North Am* 38 (3): 425–440. https://doi.org/10.1016/j.ogc.2011.05.002.

Gold, E. B., G. Block, S. Crawford, L. Lachance, G. FitzGerald, H. Miracle, and S. Sherman. 2004. "Lifestyle and demographic factors in relation to vasomotor symptoms: baseline results from the Study of Women's Health Across the Nation." *Am J Epidemiol* 159 (12): 1189–1199. https://doi.org/10.1093/aje/kwh168.

Gold, E. B., S. L. Crawford, N. E. Avis, C. J. Crandall, K. A. Matthews, L. E. Waetjen, J. S. Lee, R. Thurston, M. Vuga, and S. D. Harlow. 2013. "Factors related to age at natural menopause: longitudinal analyses from SWAN." *Am J Epidemiol* 178 (1): 70–83. https://doi.org/10.1093/aje/kws421.

Goodman, A. H., Y. T Moses, and J. L. Jones. 2019. *Race: are we so different?* Sommerset: John Wiley & Sons.

Gould, S. J., and R. C. Lewontin. 1979. "The spandrels of San Marco and the Panglossian paradigm: a critique of the adaptationist programme." *Proc R Soc Lond B* 205: 581–598. https://doi.org/10.1098/rspb.1979.0086.

Grady, D., W. Applegate, T. Bush, C. Furberg, B. Riggs, and S. B. Hulley. 1998. "Heart and Estrogen/progestin Replacement Study (HERS): design, methods, and baseline characteristics." *Control Clin Trials* 19 (4): 314–335. https://doi.org/10.1016/s0197-2456(98)00010-5.

Gravlee, C. C. 2009. "How race becomes biology: embodiment of social inequality." *Am J Phys Anthropol* 139 (1): 47–57. https://doi.org/10.1002/ajpa.20983.

"Graying Japan to Face Unprecedented Challenges." 2014. *Nippon.com*. Accessed 4 Feb. https://www.nippon.com/en/features/h00079/.

Greene, J. G. 1998. "Constructing a standard climacteric scale." *Maturitas* 29 (1): 25–31. https://doi.org/10.1016/s0378-5122(98)00025-5.

Group, The Women's Health Initiative Study. 1998. "Design of the women's health initiative clinical trial and observational study." *Control Clin Trials* 19 (1): 61–109. https://doi.org/10.1016/S0197-2456(97)00078-0.

Gultekin, E., and F. Yildiz. 2006. "Introduction to phytoestrogens." In: *Phytoestrogens in Functional Foods*, Yildiz F, (ed.). 3–18. Boca Raton, FL: CRC Press.

Gurven, M., J. Stieglitz, B. Trumble, A. D. Blackwell, B. Beheim, H. Davis, P. Hooper, and H. Kaplan. 2017. "The Tsimane Health and Life History Project: integrating anthropology and biomedicine." *Evol Anthropol* 26 (2): 54–73. https://doi.org/10.1002/evan.21515.

Hamilton, W. D. 1966. "The moulding of senescence by natural selection." *J Theoret Biol* 12 (1): 12–45. https://doi.org/10.1016/0022-5193(66)90184-6.

Harlow, S. D., S. L. Crawford, B. Sommer, and G. A. Greendale. 2000. "Self-defined menopausal status in a multi-ethnic sample of midlife women." *Maturitas* 36 (2): 93–112. https://doi.org/10.1016/s0378-5122(00)00145-6.

Harlow, S. D., M. Gass, J. E. Hall, R. Lobo, P. Maki, R. W. Rebar, S. Sherman, P. M. Sluss, and T. J. de Villiers. 2012. "Executive summary of the Stages of Reproductive Aging Workshop + 10: addressing the unfinished agenda of staging reproductive aging." *Menopause* 19 (4): 387–395. https://doi.org/10.1097/gme.0b013e31824d8f40.

Hawkes, K., J. F. O'Connell, N. G. Blurton Jones. 1997. "Hadza women's time allocation, offspring provisioning, and the evolution of long postmenopausal life spans." *Curr Anthropol.* 38(4):551–577.

Hawkes, K., J. F. O'Connell, N. G. Blurton Jones, H. Alvarez, and E. L. Charnov. 1998. "Grandmothering, menopause, and the evolution of human life histories." *Proc Natl Acad Sci.* 95(3):1336–1339. https://doi.org/10.1073/pnas.95.3.1336

Hawkes, K., and J. E. Coxworth. 2013. "Grandmothers and the evolution of human longevity: a review of findings and future directions." *Evol Anthropol Issues News Rev* 22 (6): 294–302. https://doi.org/10.1002/evan.21382.

He, C., P. Kraft, C. Chen, J. E. Buring, G. Paré, S. E. Hankinson, S. J. Chanock, P. M. Ridker, D. J. Hunter, and D. I. Chasman. 2009. "Genome-wide association studies identify loci associated with age at menarche and age at natural menopause." *Nat Genet* 41 (6): 724–728. https://doi.org/10.1038/ng.385.

He, H., B. Xie, and L. Xie. 2018. "Male pattern baldness and incidence of prostate cancer: a systematic review and meta-analysis." *Medicine (Baltimore)* 97 (28): e11379. https://doi.org/10.1097/md.0000000000011379.

Healthy People 2030. 2025. "Social determinants of health." Accessed 21 May 2025. https://odphp.health.gov/healthypeople/priority-areas/social-determinants-health.

Hegsted, D. M. 2001. "Fractures, calcium, and the modern diet." *Am J Clin Nutr* 74 (5): 571–573. https://doi.org/10.1093/ajcn/74.5.571.

Henrich, J., S. J. Heine, and A. Norenzayan. 2010. "The weirdest people in the world?" *Behav Brain Sci* 33 (2–3): 61–83. https://doi.org/10.1017/S0140525X0999152X.

Henry, L. P., M. Bruijning, S. K. G. Forsberg, and J. F. Ayroles. 2021. "The microbiome extends host evolutionary potential." *Nat Commun* 12 (1): 5141. https://doi.org/10.1038/s41467-021-25315-x.

Hilliard, C. B. 2016. "High osteoporosis risk among East Africans linked to lactase persistence genotype." *Bonekey Rep* 5: 803. https://doi.org/10.1038/bonekey.2016.30.

Hofstede, G. 2011. "Dimensionalizing cultures: the Hofstede model in context." *Online Read Psychol Cult* 2, no. 1: 1–26. https://doi.org/10.9707/2307-0919.1014.

Howard, B. 2014. *Andropause: the complete male menopause guide.* Elsevier: Jett Publishing.

Howick, J., P. Kelly, and M. Kelly. 2019. "Establishing a causal link between social relationships and health using the Bradford Hill Guidelines." *SSM Popul Health* 8: 100402. https://doi.org/10.1016/j.ssmph.2019.100402.

Hrdy, S. B. 2009. *Mothers and others: the evolutionary origins of mutual understanding.* Cambridge, MA Belknap Press of Harvard University Press. 422 pages.

Hulley, S., D. Grady, T. Bush, C. Furberg, D. Herrington, B. Riggs, and E. Vittinghoff. 1998. "Randomized trial of estrogen plus progestin for secondary prevention of coronary heart disease in postmenopausal women. Heart and Estrogen/progestin Replacement Study (HERS) research group." *JAMA* 280 (7): 605–613. https://doi.org/10.1001/jama.280.7.605.

Hvas, L. 2006. "Menopausal women's positive experience of growing older." *Maturitas* 54 (3): 245–251. https://doi.org/10.1016/j.maturitas.2005.11.006.

Iino, C., T. Shimoyama, K. Iino, Y. Yokoyama, D. Chinda, H. Sakuraba, S. Fukuda, and S. Nakaji. 2019. "Daidzein intake is associated with equol producing status through an increase in the intestinal bacteria responsible for equol production." *Nutrients* 11 (2):1–22. https://doi.org/10.3390/nu11020433.

Jablonski, N. G. 2021. "Skin color and race." *Am J Phys Anthropol* 175 (2): 437–447. https://doi.org/10.1002/ajpa.24200.

Jackson, F. L. C. 1991. "Secondary compounds in plants (allelochemicals) as promoters of human biological variability." *Annu Rev Anthropol* 20: 505–546.

"Japan's population projected to fall to 87 million in 2070." 2023. *Nippon.com*. Accessed 22 May 2025. https://www.nippon.com/en/japan-data/h01664/.

Japanese Ministry of Health 1996. *Kokumin eiyoo no genjoo - Heisei rokunen kokumin eiyoo choosa seiseki. (The present state of Nutrition in Japan: The report on the 1994 National Nutrition Survey)*. Japanese Ministry of Health (Tokyo).

Jermyn, D. "Whose menopause revolution? Investigating the UK's 'Davina effect' and the contemporary menopause market." *Eur J Cult Stud* 0 (0): 1–19. https://doi.org/10.1177/13675494241287931.

Jin, K. 2010. "Modern biological theories of aging." *Aging Dis* 1 (2): 72–74.

Johns, T. 1996. *The origins of human diet and medicine: chemical ecology*. Tucson: University of Arizona Press.

Johnson, A., L. Roberts, and G. Elkins. 2019. "Complementary and alternative medicine for menopause." *J Evid Based Integr Med* 24: 2515690x19829380. https://doi.org/10.1177/2515690x19829380.

Jones, J. M., M. Kohli, and C. L. Loprinzi. 2012. "Androgen deprivation therapy-associated vasomotor symptoms." *Asian J Androl* 14 (2): 193–197. https://doi.org/10.1038/aja.2011.101.

Kaufert, P. A. 1982. "Anthropology and the menopause: the development of a theoretical framework." *Maturitas* 4 (3): 181–193. https://doi.org/10.1016/0378-5122(82)90048-2.

Kirkwood, T. B. L. 1977. "Evolution of ageing." *Nature* 270 (5635): 301–304.

Kirkwood, T. B. L. 2005. "Understanding the odd science of aging." *Cell* 120 (4): 437–447. https://doi.org/10.1016/j.cell.2005.01.027.

Kirkwood, T. L., P. Kapahi, and D. P. Shanley. 2000. "Evolution, stress, and longevity." *J Anat* 197 (Pt 4): 587–590. https://doi.org/10.1046/j.1469-7580.2000.19740587.x.

Kirmayer, L. J., and A. Young. 1998. "Culture and somatization: clinical, epidemiological, and ethnographic perspectives." *Psychosom Med* 60 (4): 420–430. https://doi.org/10.1097/00006842-199807000-00006.

Kleinman, A. M. 1977. "Depression, somatization and the "new cross-cultural psychiatry"." *Soc Sci Med (1967)* 11 (1): 3–9. https://doi.org/10.1016/0037-7856(77)90138-X.

Koettl, J. 2015. "Did we get the 'old-age dependency' of aging countries all wrong?". *Brookings*. Accessed 22 May 2025. https://www.brookings.edu/articles/did-we-get-the-old-age-dependency-of-aging-countries-all-wrong/#:~:text=The%20old%2Dage%20dependency%20ratio,retirement%20have%20begun%20to%20shift.

Kohn, G. E., K. M. Rodriguez, J. Hotaling, and A. W. Pastuszak. 2019. "The history of estrogen therapy." *Sex Med Rev* 7 (3): 416–421. https://doi.org/10.1016/j.sxmr.2019.03.006.

Koike, Y. 2015. "Why inequality is different in Japan." *World Economic Forum*. Accessed 27 May2025 https://www.weforum.org/stories/2015/03/why-inequality-is-different-in-japan/.

Lampe, J. W. 2009. "Is equol the key to the efficacy of soy foods?" *Am J Clin Nutr* 89 (5): 1664s–1667s. https://doi.org/10.3945/ajcn.2009.26736T.

Le Couteur, D. G., and J. Thillainadesan. 2022. "What is an aging-related disease? An epidemiological perspective." *J Gerontol A Biol Sci Med Sci* 77 (11): 2168–2174. https://doi.org/10.1093/gerona/glac039.

Levine, M. E., A. T. Lu, A. Quach, B. H. Chen, T. L. Assimes, S. Bandinelli, L. Hou, A. A. Baccarelli, J. D. Stewart, Y. Li, E. A. Whitsel, J. G. Wilson, A. P. Reiner, A. Aviv, K. Lohman, Y. Liu, L. Ferrucci, and S. Horvath. 2018. "An epigenetic biomarker of aging for lifespan and healthspan." *Aging (Albany NY)* 10 (4): 573–591. https://doi.org/10.18632/aging.101414.

Levy, N., J. Sturgess, and P. Mills. 2018. ""Pain as the fifth vital sign" and dependence on the "numerical pain scale" is being abandoned in the US: why?" *Br J Anaesth* 120 (3): 435–438. https://doi.org/10.1016/j.bja.2017.11.098.

Lewis, C., P. R. Cohen, D. Bahl, E. M. Levine, and W. Khaliq. 2023. "Race and ethnic categories: a brief review of global terms and nomenclature." *Cureus* 15 (7): e41253. https://doi.org/10.7759/cureus.41253.

Lind, M. I., and F. Spagopoulou. 2018. "Evolutionary consequences of epigenetic inheritance." *Heredity* 121 (3): 205–209. https://doi.org/10.1038/s41437-018-0113-y.

Littler, J. 2019. "The art of perseverance: how gaman defined Japan." *BBC*. Accessed 27 May 2025. https://www.bbc.com/worklife/article/20190319-the-art-of-perseverance-how-gaman-defined-japan.

Lock, M. 1980. *East Asian medicine in urban Japan*. Edited by Charles Leslie. *Comparative Studies of Health Systems and Medical Care*. Berkeley: University of California Press.

Lock, M. 1993. *Encounters with aging: mythologies of menopause in Japan and North America*. Berkeley: University of California Press.

Lock, M. 2017. "Recovering the body." *Annu Rev Anthropol* 46 (2017): 1–14. https://doi.org/10.1146/annurev-anthro-102116-041253.

Lock, M. M. 1984. "East Asian medicine and health care for the Japanese elderly." *Pac Aff* 57 (1): 65–73. https://doi.org/10.2307/2758386.

Lock, M., and P. Kaufert. 2001. "Menopause, local biologies, and cultures of aging." *Am J Hum Biol* 13 (4): 494–504. https://doi.org/10.1002/ajhb.1081.

Lotufo, P. A., C. U. Chae, U. A. Ajani, C. H. Hennekens, and J. E. Manson. 2000. "Male pattern baldness and coronary heart disease: the Physicians' Health Study." *Arch Intern Med* 160 (2): 165–171. https://doi.org/10.1001/archinte.160.2.165.

Lu, J., Y. Shin, M. S. Yen, and S. S. Sun. 2016. "Peak bone mass and patterns of change in total bone mineral density and bone mineral contents from childhood into young adulthood." *J Clin Densitom* 19 (2): 180–191. https://doi.org/10.1016/j.jocd.2014.08.001.

Luhrmann, T. M. 2001. "Thick description: methodology." In *International encyclopedia of the social & behavioral sciences*, edited by N. J. Smelser and P. B. Baltes, 15665–15668. Oxford: Pergamon.

Maeda, Y., K. Takahashi, K. Yamamoto, T. Tanimoto, M. Kami, and A. Crump. 2019. "Factors affecting the provision of analgesia during childbirth, Japan." *Bull World Health Organ* 97 (9): 631–636. https://doi.org/10.2471/blt.19.230128.

Maklakov, A. A., and T. Chapman. 2019. "Evolution of ageing as a tangle of trade-offs: energy versus function." *Proc Biol Sci* 286 (1911): 20191604. https://doi.org/10.1098/rspb.2019.1604.

Marlowe, F. 2000. “The patriarch hypothesis: an alternative explanation of menopause.” *Hum Nat* 11 (1): 27–42. https://doi.org/10.1007/s12110-000-1001-7.

Marmot, M. G., and S. L. Syme. 1976. “Acculturation and coronary heart disease in Japanese-Americans.” *Am J Epidemiol* 104 (3): 225–247. https://doi.org/10.1093/oxfordjournals.aje.a112296.

Martin, E. 1991. “The egg and the sperm: how science has constructed a romance based on stereotypical male-female roles.” *Signs* 16 (3): 485–501. https://www.jstor.org/stable/3174586.

Martinez-Radl, F. B., D. E. Hinton, and U. Stangier. 2023. “Susto as a cultural conceptualization of distress: existing research and aspects to consider for future investigations.” *Transcult Psychiatry* 60 (4): 690–702. https://doi.org/10.1177/13634615231163986.

Mayo, B., L. Vázquez, and A. B. Flórez. 2019. “Equol: a bacterial metabolite from the daidzein isoflavone and its presumed beneficial health effects.” *Nutrients* 11 (9): 2231. https://doi.org/10.3390/nu11092231.

McAuley, M. T. 2023. “An evolutionary perspective of lifespan and epigenetic inheritance.” *Exp Gerontol* 179: 112256. https://doi.org/10.1016/j.exger.2023.112256.

McHugh, D., and J. Gil. 2018. “Senescence and aging: causes, consequences, and therapeutic avenues.” *J Cell Biol* 217 (1): 65–77. https://doi.org/10.1083/jcb.201708092.

Medawar, P. B. 1952. *An unsolved problem of biology: an inaugural lecture delivered at University College, London, 6 December, 1951*. H.K. Lewis and Company.

Melby, M. K. 2005a. “Factor analysis of climacteric symptoms in Japan.” *Maturitas* 52 (3–4): 205–222. https://doi.org/10.1016/j.maturitas.2005.02.002.

Melby, M. K. 2005b. “Vasomotor symptom prevalence and language of menopause in Japan.” *Menopause* 12 (3): 250–257. https://journals.lww.com/menopausejournal/abstract/2005/12030/vasomotor_symptom_prevalence_and_language_of.5.aspx

Melby, M. K. 2006. “Climacteric symptoms among Japanese women and men: comparison of four symptom checklists.” *Climacteric* 9: 298–304. https://doi.org/10.1080/13697130600868653.

Melby, M. K. 2007a. “Biocultural approaches to the study of the local biology of menopause in Japan.” Ph.D. Dissertation, Emory University

Melby, M. K. 2007b. “Chilliness: a vasomotor symptom in Japan.” *Menopause* 14 (4): 1–8. https://doi.org/10.1097/gme.0b013e31804ffd81.

Melby, M. K. 2016. “Lessons on aging: hopes and concerns of Japanese women at midlife.” *J Women Aging* 28 (2): 127–140. https://doi.org/10.1080/08952841.2014.950544.

Melby, M. K., and M. Lampl. 2011. “Menopause, A biocultural perspective.” *Annu Rev Anthropol* 40 (1): 53–70. https://doi.org/10.1146/annurev-anthro-081309-145641.

Melby, M. K., M. Lock, and P. Kaufert. 2005. “Culture and symptom reporting at menopause.” *Hum Reprod Update* 11 (5): 495–512. https://doi.org/10.1093/humupd/dmi018.

Melby, M. K., 裕太 石川, and 国比古 富永. 2015. “ベジタリアンは長寿か—アンチエイジング（抗加齢） 医学からの検証 (Vegetarianism and longevity —examination from the perspective of anti-aging medicine).” In *アンチエイジング医学の基礎と臨床 (Anti-aging medicine: foundation and clinical)*, edited by Japanese

Board of Anti-Aging Medicine Specialists and Health Practitioners (3rd edition), pp. 243–244. Tokyo, Japan: Medical View Co.

Melby, M. K., S. Watanabe, P. L. Whitten, and C. M. Worthman. 2005. "Sensitive high-performance liquid chromatographic method using coulometric electrode array detection for measurement of phytoestrogens in dried blood spots." *J Chromatogr B Analyt Technol Biomed Life Sci* 826 (1–2): 81–90. https://doi.org/10.1016/j.jchromb.2005.08.009.

Melby, M. K., L. L. Sievert, D. Anderson, and C. M. Obermeyer. 2011. "Overview of methods used in cross-cultural comparisons of menopausal symptoms and their determinants: guidelines for Strengthening the Reporting of Menopause and Aging (STROMA) studies." *Maturitas* 70 (2): 99–109. https://doi.org/10.1016/j.maturitas.2011.07.011.

Melby, M. K., E. Zent, S. Ariste, R. Shoukat, and M. Nichter. 2025. "'The Littlest Creatures that Live Inside Us': Public understandings influencing microbiome-related behaviors." *Social Science & Medicine*, 376:117864. https://doi.org/10.1016/j.socscimed.2025.117864

Meng, L., Y. Zhang, Y. Hua, Y. Ma, H. Wang, X. Li, Y. Jiang, and G. Zhu. 2023. "Identification of oogonial stem cells in chicken ovary." *Cell Prolif* 56 (3): e13371. https://doi.org/10.1111/cpr.13371.

Messina, M., A. Duncan, V. Messina, H. Lynch, J. Kiel, and J. W. Erdman. 2022. "The health effects of soy: a reference guide for health professionals." *Front Nutr* 9: 970364. https://doi.org/10.3389/fnut.2022.970364.

Messina, M., and L. Hilakivi-Clarke. 2009. "Early intake appears to be the key to the proposed protective effects of soy intake against breast cancer." *Nutr Cancer* 61 (6): 792–798. https://doi.org/10.1080/01635580903285015.

Miller, G. (host). 2009 (March 17). "Menopause and medicine. Louise Foxcroft: Hot Flushes, Cold Science." (No 25) [Audio podcast episode]. In *Podularity* Podcast. https://podularity.com/2009/03/17/25-menopause-and-medicine/

Moerman, D. E. 2002. *Meaning, medicine and the 'placebo effect'*. Edited by Alan Harwood. *Cambridge Studies in Medical Anthropology*. Cambridge: Cambridge University Press.

Monteleone, P., G. Mascagni, A. Giannini, A. R. Genazzani, and T. Simoncini. 2018. "Symptoms of menopause - global prevalence, physiology and implications." *Nat Rev Endocrinol* 14 (4): 199–215. https://doi.org/10.1038/nrendo.2017.180.

Moqri, M., C. Herzog, J. R. Poganik, J. Justice, D. W. Belsky, A. Higgins-Chen, A. Moskalev, G. Fuellen, A. A. Cohen, I. Bautmans, M. Widschwendter, J. Ding, A. Fleming, J. Mannick, J. J. Han, A. Zhavoronkov, N. Barzilai, M. Kaeberlein, S. Cummings, B. K. Kennedy, L. Ferrucci, S. Horvath, E. Verdin, A. B. Maier, M. P. Snyder, V. Sebastiano, and V. N. Gladyshev. 2023. "Biomarkers of aging for the identification and evaluation of longevity interventions." *Cell* 186 (18): 3758–3775. https://doi.org/10.1016/j.cell.2023.08.003.

Morito, K., T. Hirose, J. Kinjo, T. Hirakawa, M. Okawa, T. Nohara, S. Ogawa, S. Inoue, M. Muramatsu, and Y. Masamune. 2001. "Interaction of phytoestrogens with estrogen receptors alpha and beta." *Biol Pharm Bull* 24 (4): 351–356. https://doi.org/10.1248/bpb.24.351.

Morton, R. A., J. R. Stone, and R. S. Singh. 2013. "Mate choice and the origin of menopause." *PLoS Comput Biol* 9 (6): e1003092. https://doi.org/10.1371/journal.pcbi.1003092.

"Most prolific mother ever." 2025. *Guinness World Records*. Accessed 8 Feb. https://www.guinnessworldrecords.com/world-records/most-prolific-mother-ever.

Nagata, C. 2021. "Soy intake and chronic disease risk: findings from prospective cohort studies in Japan." *Eur J Clin Nutr* 75 (6): 890–901. https://doi.org/10.1038/s41430-020-00744-x.

Nagata, C., H. Shimizu, R. Takami, M. Hayashi, N. Takeda, and K. Yasuda. 1999. "Hot flushes and other menopausal symptoms in relation to soy product intake in Japanese women." *Climacteric*. 2:6–12. https://doi.org/10.3109/13697139909025557

NAMS Advisory Panel. 2022. "The 2022 hormone therapy position statement of The North American Menopause Society." *Menopause* 29 (7): 767–794. https://doi.org/10.1097/gme.0000000000002028.

Narasimhan, H., C. C. Ren, S. Deshpande, and K. E. Sylvia. 2021. "Young at gut-turning back the clock with the gut microbiome." *Microorganisms* 9 (3): 555. https://doi.org/10.3390/microorganisms9030555.

Narzary, D. C. 2004. "The myths of Japanese 'homogeneity'." *China Rep* 40 (3): 311–319. https://doi.org/10.1177/000944550404000308.

Nesse, R. M., and G. C. Williams. 1994. *Why we get sick: the new science of Darwinian medicine*. 1st ed. New York: Times Books.

Newton, K. M., S. D. Reed, S. Uchiyama, C. Qu, T. Ueno, S. Iwashita, G. Gunderson, S. Fuller, and J. W. Lampe. 2015. "A cross-sectional study of equol producer status and self-reported vasomotor symptoms." *Menopause* 22 (5): 489–495. https://doi.org/10.1097/GME.0000000000000363.

Niewöhner, J., and M. Lock. 2018. "Situating local biologies: anthropological perspectives on environment/human entanglements." *BioSocieties* 13 (4): 681–697. https://doi.org/10.1057/s41292-017-0089-5.

Noel, S. E., M. P. Santos, and N. C. Wright. 2021. "Racial and ethnic disparities in bone health and outcomes in the United States." *J Bone Miner Res* 36 (10): 1881–1905. https://doi.org/10.1002/jbmr.4417.

O'Toole, P. W., and I. B. Jeffery. 2015. "Gut microbiota and aging." *Science* 350 (6265): 1214–1215. https://doi.org/10.1126/science.aac8469.

OECD. 2019. "Health at a glance 2019: OECD indicators." OECD Publishing. Accessed 4 Feb. https://doi.org/10.1787/4dd50c09-en.

Ortiz, C., and B. Manta. 2024. "Advances in equol production: sustainable strategies for unlocking soy isoflavone benefits." *Res Chem* 7: 101288. https://doi.org/10.1016/j.rechem.2023.101288.

Oseni, T., R. Patel, J. Pyle, and V. C. Jordan. 2008. "Selective estrogen receptor modulators and phytoestrogens." *Planta Med* 74 (13): 1656–1665. https://doi.org/10.1055/s-0028- 1088304.

Pachman, D. R., C. L. Loprinzi, P. J. Novotny, D. V. Satele, B. M. Linquist, S. Wolf, and D. L. Barton. 2013. "Sternal skin conductance: a reasonable surrogate for hot flash measurement?" *Menopause* 20 (11): 1164–1168. https://doi.org/10.1097/GME.0b013e31828ccc53.

Pal, S., and J. K. Tyler. 2016. "Epigenetics and aging." *Sci Adv* 2 (7): e1600584. https://doi.org/10.1126/sciadv.1600584.

Pamplona, R., M. Jové, J. Gómez, and G. Barja. 2023. "Programmed versus non-programmed evolution of aging. What is the evidence?" *Exp Gerontol* 175: 112162. https://doi.org/10.1016/j.exger.2023.112162.

Park, J., and M. E. Johantgen. 2017. "A cross-cultural comparison of symptom reporting and symptom clusters in heart failure." *J Transcult Nurs* 28 (4): 372–380. https://doi.org/10.1177/1043659616651673.

Parker, A., S. Romano, R. Ansorge, A. Aboelnour, G. Le Gall, G. M. Savva, M. G. Pontifex, A. Telatin, D. Baker, E. Jones, D. Vauzour, S. Rudder, L. A. Blackshaw, G. Jeffery, and S. R. Carding. 2022. "Fecal microbiota transfer between young and aged mice reverses hallmarks of the aging gut, eye, and brain." *Microbiome* 10 (1): 68. https://doi.org/10.1186/s40168-022-01243-w.

Passarino, G., F. De Rango, and A. Montesanto. 2016. "Human longevity: genetics or lifestyle? It takes two to tango." *Immun Ageing* 13 (1): 12. https://doi.org/10.1186/s12979-016-0066-z.

Payer, L. 1988. *Medicine & culture: varieties of treatment in the United States, England, West Germany, and France*. New York: Henry Holt and Co.

Penn, D. J., and S. Számadó. 2020. "The Handicap Principle: how an erroneous hypothesis became a scientific principle." *Biol Rev Camb Philos Soc* 95 (1): 267–290. https://doi.org/10.1111/brv.12563.

Perez, M. F., and B. Lehner. 2019. "Intergenerational and transgenerational epigenetic inheritance in animals." *Nat Cell Biol* 21 (2): 143–151. https://doi.org/10.1038/s41556-018-0242-9.

Perls, T. T., and R. C. Fretts. 2001. "The evolution of menopause and human life span." *Ann Hum Biol* 28 (3): 237–245. https://doi.org/10.1080/030144601300119052.

Pifferi, F., and F. Aujard. 2019. "Caloric restriction, longevity and aging: recent contributions from human and non-human primate studies." *Prog Neuropsychopharmacol Biol Psychiatry* 95: 109702.

Pike, K. L. 1954. *Language in relation to a unified theory of the structure of human behavior. Language in relation to a unified theory of the structure of human behavior.* Dallas, TX: Summer Institute of Linguistics.

Pimpton, G. 1990. "Maya Angelou, the art of fiction No. 119." *Paris Rev* Fall 1990 (116): 144–167.

Pines, A. 2011. "Male menopause: is it a real clinical syndrome?" *Climacteric* 14 (1): 15–17. https://doi.org/10.3109/13697137.2010.507442.

"Podularity." 2009. In *Menopause and medicine. Louise Foxcroft: Hot flushes, cold science*, 19:07, https://podularity.com/2009/03/17/25-menopause-and-medicine/.

Poulain, M., G. M. Pes, C. Grasland, C. Carru, L. Ferrucci, G. Baggio, C. Franceschi, and L. Deiana. 2004. "Identification of a geographic area characterized by extreme longevity in the Sardinia island: the AKEA study." *Exp Gerontol* 39 (9): 1423–1429. https://doi.org/10.1016/j.exger.2004.06.016.

Prestinaci, F., P. Pezzotti, and A. Pantosti. 2015. "Antimicrobial resistance: a global multifaceted phenomenon." *Pathog Glob Health* 109 (7): 309–318. https://doi.org/10.1179/2047773215y.0000000030.

Prior, J. C. 2011. "Progesterone for symptomatic perimenopause treatment - progesterone politics, physiology and potential for perimenopause." *Facts Views Vis Obgyn* 3 (2): 109–120. https://pmc.ncbi.nlm.nih.gov/articles/PMC3987489/.

Prior, J. 2021. "Is estrogen deficiency really a thing?" *Scientific American*. https://www.scientificamerican.com/article/is-estrogen-deficiency-really-a-thing/.

Prior, J. C. 2022. "Adaptive, reversible, hypothalamic reproductive suppression: more than functional hypothalamic amenorrhea." *Front Endocrinol (Lausanne)* 13: 893889. https://doi.org/10.3389/fendo.2022.893889.

Provenzano, F., and M. Deleidi. 2021. "Reassessing neurodegenerative disease: immune protection pathways and antagonistic pleiotropy." *Trends Neurosci* 44 (10): 771–780. https://doi.org/10.1016/j.tins.2021.06.006.

Prum, R. O. 2017. *The evolution of beauty: how Darwin's forgotten theory of mate choice shapes the animal world - and us*. New York: Doubleday.

Rafii, F. 2015. "The role of colonic bacteria in the metabolism of the natural isoflavone daidzin to equol." *Metabolites* 5 (1): 56–73. https://doi.org/10.3390/metabo5010056.

Räsänen, J. 2021. "Does the distinction between biological and chronological age support legal age change?". *News*. Last Modified 1 Apr 2021. Accessed 8 Jan. https://bioethics.hms.harvard.edu/journal/legal-age-change#_edn1.

Ríos-Covian, D., P. Langella, and R. Martin. 2021. "From short- to long-term effects of C-section delivery on microbiome establishment and host health." *Microorganisms* 9 (10): 2122. https://doi.org/10.3390/microorganisms9102122.

Rosenberg, K. R., and W. R. Trevathan. 2021. "The obstetrical dilemma revisited - revisited." In *The Routledge handbook of anthropology and reproduction*, Han, S., and Tomori, C. (eds). 425–440. London: Routledge.

Rosenberger, N. R. 1992. "The process of discourse: usages of a Japanese medical term." *Soc Sci Med* 34 (3): 237–247. https://doi.org/10.1016/0277-9536(92)90266-S.

Rossouw, J. E., G. L. Anderson, R. L. Prentice, A. Z. LaCroix, C. Kooperberg, M. L. Stefanick, R. D. Jackson, S. A. Beresford, B. V. Howard, K. C. Johnson, J. M. Kotchen, and J. Ockene. 2002. "Risks and benefits of estrogen plus progestin in healthy postmenopausal women: principal results from the Women's Health Initiative randomized controlled trial." *JAMA* 288 (3): 321–333. https://doi.org/10.1001/jama.288.3.321.

Saunders, T. 2024. "The oldest person in the world in 2025." The Human Body. *BBC*. Accessed 8 Jan. https://www.sciencefocus.com/the-human-body/oldest-person-in-the-world.

Schug, T. T., A. F. Johnson, L. S. Birnbaum, T. Colborn, L. J. Guillette, Jr., D. P. Crews, T. Collins, A. M. Soto, F. S. vom Saal, J. A. McLachlan, C. Sonnenschein, and J. J. Heindel. 2016. "Minireview: endocrine disruptors: past lessons and future directions." *Mol Endocrinol* 30 (8): 833–847. https://doi.org/10.1210/me.2016-1096.

Schulster, M., A. M. Bernie, and R. Ramasamy. 2016. "The role of estradiol in male reproductive function." *Asian J Androl* 18 (3): 435–440. https://doi.org/10.4103/1008- 682x.173932.

Schwarz, K. G., S. C. Vicencio, N. C. Inestrosa, P. Villaseca, and R. Del Rio. 2024. "Autonomic nervous system dysfunction throughout menopausal transition: a potential mechanism underpinning cardiovascular and cognitive alterations during female ageing." *J Physiol* 602 (2): 263–280. https://doi.org/10.1113/JP285126. https://physoc.onlinelibrary.wiley.com/doi/abs/10.1113/JP285126.

Sear, R., and R. Mace. 2008. "Who keeps children alive? A review of the effects of kin on child survival." *Evol Human Behavior* 29 (1): 1–18. https://doi.org/10.1016/j.evolhumbehav.2007.10.001.

Sender, R., S. Fuchs, and R. Milo. 2016. "Revised estimates for the number of human and bacteria cells in the body." *PLoS Biol* 14 (8): e1002533. https://doi.org/10.1371/journal.pbio.1002533.

Setchell, K. D. 2001. "Soy isoflavones--benefits and risks from nature's selective estrogen receptor modulators (SERMs)." *J Am Coll Nutr* 20 (5 Suppl): 354S–362S; discussion 381S–383S. https://doi.org/10.1080/07315724.2001.10719168.

Setchell, K. D. R., N. M. Brown, and E. Lydeking-Olsen. 2002. "The clinical importance of the metabolite equol—a clue to the effectiveness of soy and its isoflavones." *J Nutr* 132 (12): 3577–3584. https://doi.org/10.1093/jn/132.12.3577.

Shay, J. W. 2016. "Role of telomeres and telomerase in aging and cancer." *Cancer Discov* 6 (6): 584–593. https://doi.org/10.1158/2159-8290.Cd-16-0062.

Shlisky, J., R. Mandlik, S. Askari, S. Abrams, J. M. Belizan, M. W. Bourassa, G. Cormick, A. Driller-Colangelo, F. Gomes, A. Khadilkar, V. Owino, J. M. Pettifor, Z. H. Rana, D. E. Roth, and C. Weaver. 2022. "Calcium deficiency worldwide: prevalence of inadequate intakes and associated health outcomes." *Ann N Y Acad Sci* 1512 (1): 10–28. https://doi.org/10.1111/nyas.14758.

Short, R. V. 1976. "The evolution of human reproduction." *Proc R Soc Lond B Biol Sci* 195 (1118): 3–24. https://doi.org/10.1098/rspb.1976.0095.

Shufelt, C. L. 2025. "Hormone therapy." The Menopause Society. Accessed 27 May 2025. https://menopause.org/patient-education/menopause-topics/hormone-therapy.

Sievert, L. L. 2006. *Menopause: A Biocultural Perspective.Studies in medical anthropology*. New Brunswick, NJ: Rutgers University Press.

Sievert, L. L., A. Reza, P. Mills, L. Morrison, N. Rahberg, A. Goodloe, M. Sutherland, and D. E. Brown. 2010. "Diurnal rhythm and concordance between objective and subjective hot flashes: the Hilo Women's Health Study." *Menopause* 17 (3): 471–479. https://doi.org/10.1097/gme.0b013e3181cbb3c6.

Skloot, R. 2010. *The immortal life of Henrietta Lacks*. New York: Crown Publishers.

Smith, P, D. Willemsen, M. Popkes, F. Metge, E. Gandiwa, M. Reichard, and D. R. Valenzano. 2017. "Regulation of life span by the gut microbiota in the short-lived African turquoise killifish." *eLife* 6: e27014. https://doi.org/10.7554/eLife.27014.

Spadaro, O., Y. Youm, I. Shchukina, S. Ryu, S. Sidorov, A. Ravussin, K. Nguyen, E. Aladyeva, A. N. Predeus, S. R. Smith, E. Ravussin, C. Galban, M. N. Artyomov, and V. D. Dixit. 2022. "Caloric restriction in humans reveals immunometabolic regulators of health span." *Science* 375 (6581): 671–677. https://doi.org/10.1126/science.abg7292.

Stearns, F. W. 2010. "One hundred years of pleiotropy: a retrospective." *Genetics* 186 (3): 767–773. https://doi.org/10.1534/genetics.110.122549.

Stearns, S. C. 1992. *The evolution of life histories*. New York: Oxford University Press.

Stenvinkel, P., and P. G. Shiels. 2019. "Long-lived animals with negligible senescence: clues for ageing research." *Biochem Soc Trans* 47 (4): 1157–1164. https://doi.org/10.1042/bst20190105.

Stordal, B. 2023. "Breastfeeding reduces the risk of breast cancer: a call for action in high-income countries with low rates of breastfeeding." *Cancer Med* 12 (4): 4616–4625. https://doi.org/10.1002/cam4.5288.

Sweatt, K., W. Timothy Garvey, and C. Martins. 2024. "Strengths and limitations of BMI in the diagnosis of obesity: what is the path forward?" *Curr Obes Rep* 13 (3): 584–595. https://doi.org/10.1007/s13679-024-00580-1.

Taku, K., M. K. Melby, F. Kronenberg, M. S. Kurzer, and M. Messina. 2012. "Extracted or synthesized soybean isoflavones reduce menopausal hot flash frequency and severity: systematic review and meta-analysis of randomized controlled trials." *Menopause* 19 (7): 776–790. https://doi.org/10.1097/gme.0b013e3182410159.

Tamashiro, T. 2018. "How to Ikigai." *TEDxYYC*. Last Modified 8 Sep. Accessed 9 Feb. https://www.youtube.com/watch?v=pk-PcJS2QaU.

Trindade LS, Aigaki T, Peixoto AA, Balduino A, Manica da Cruz IB, Heddle JG. 2013. A novel classification system for evolutionary aging theories. Front Genet. 4: 1–8. https://doi.org/10.3389/fgene.2013.00025

Trindade, L.S., T. Aigaki, A.A. Peixoto, A. Balduino, I.B. Manica da Cruz, J.G. Heddle. 2013. "A novel classification system for evolutionary aging theories." *Front Genet*. 4: 1-8. https://doi.org/10.3389/fgene.2013.00025

Troisi, A. 2020. "Social stress and psychiatric disorders: evolutionary reflections on debated questions." *Neurosci Biobehav Rev* 116: 461–469. https://doi.org/10.1016/j.neubiorev.2020.07.016.

Tuljapurkar, S. D., C. O. Puleston, and M. D. Gurven. 2007. "Why men matter: mating patterns drive evolution of human lifespan." *PLoS One* 2 (8): e785. https://doi.org/10.1371/journal.pone.0000785.

UC Museum of Paleontology. Evolution 101. *Understanding Evolution*. Available at: https://evolution.berkeley.edu/evolution-101/. Accessed 9 Feb 25, 2025.

UN. 2022. "World population prospects 2022: summary of results." Population Division. United Nations. Accessed 27 May 2025. https://www.un.org/development/desa/pd/content/World-Population-Prospects-2022.

van den Beld, A. W., J. M. Kaufman, M. C. Zillikens, S. W. J. Lamberts, J. M. Egan, and A. J. van der Lely. 2018. "The physiology of endocrine systems with ageing." *Lancet Diabetes Endocrinol* 6 (8): 647–658. https://doi.org/10.1016/s2213-8587(18)30026-3.

Vance, D. A. D. 2007. "Premarin: the intriguing history of a controverisal drug." *Int J Pharm Compd* 11 (4): 282–286.

Vitzthum, V. J. 2009. "The ecology and evolutionary endocrinology of reproduction in the human female." *Am J Phys Anthropol* 140 (Suppl 49): 95–136. https://doi.org/10.1002/ajpa.21195.

Wade, N. 2014. "Diet's link to longevity: after 2 studies diverge, a search for consensus." 1 Apr. https://www.nytimes.com/2014/04/02/science/a-new-salvo-in-debate-over-caloric-restrictions-and-longevity.html.

Walker, J. J., J. R. Terry, K. Tsaneva-Atanasova, S. P. Armstrong, C. A. McArdle, and S. L. Lightman. 2010. "Encoding and decoding mechanisms of pulsatile hormone secretion." *J Neuroendocrinol* 22 (12): 1226–1238. https://doi.org/10.1111/j.1365-2826.2010.02087.x.

Wallace, W. H. B., and T. W. Kelsey. 2010. "Human ovarian reserve from conception to the menopause." *PLoS One* 5 (1): e8772. https://doi.org/10.1371/journal.pone.0008772.

Wang, K., H. Liu, Q. Hu, L. Wang, J. Liu, Z. Zheng, W. Zhang, J. Ren, F. Zhu, and G.-H. Liu. 2022. "Epigenetic regulation of aging: implications for interventions of aging and diseases." *Signal Transduct Targeted Ther* 7 (1): 374. https://doi.org/10.1038/s41392-022-01211-8.

Wei, Y., J. Lv, Y. Guo, Z. Bian, M. Gao, H. Du, L. Yang, Y. Chen, X. Zhang, T. Wang, J. Chen, Z. Chen, C. Yu, D. Huo, and L. Li. 2020. "Soy intake and breast cancer risk: a prospective study of 300,000 Chinese women and a dose-response meta-analysis." *Eur J Epidemiol* 35 (6): 567–578. https://doi.org/10.1007/s10654-019-00585-4.

Weon, B. M., and J. H. Je. 2009. "Theoretical estimation of maximum human lifespan." *Biogerontology* 10 (1): 65–71. https://doi.org/10.1007/s10522-008-9156-4.

WHO. 2020. "Life expectancy and health life expectancy data by country." Accessed 4 Feb. https://apps.who.int/gho/data/node.main.688.

WHO. 2024. Life expectancy at birth. Edited by WHO. https://data.who.int/indicators/i/A21CFC2/90E2E48.

Whyte, H. E. A., S. M. Findlay, Canadian Paediatric Society, and Adolescent Health Committee. 2004. "Dieting in adolescence." *Paediatr Child Health* 9 (7): 487–503. https://doi.org/10.1093/pch/9.7.487.

Wiley, A. S. 2016. *Re-imagining milk*. New York: Routledge.

Wiley, A. S., and J. M. Cullin. 2016. "What do anthropologists mean when they use the term biocultural?" *Am Anthropol* 118 (3): 554–569. https://doi.org/10.1111/aman.12608.

Willcox, D. C., B. J. Willcox, W. C. Hsueh, and M. Suzuki. 2006. "Genetic determinants of exceptional human longevity: insights from the Okinawa Centenarian Study." *Age (Dordr)* 28 (4): 313–332. https://doi.org/10.1007/s11357-006-9020-x.

Williams, G. C. 2001. "Pleiotropy, natural selection, and the evolution of senescence." *Sci Aging Knowl Environ* 2001 (1): cp13–cp13. https://doi.org/10.1126/sageke.2001.1.cp13.

Wilson, R. A. 1966. *Feminine forever*. Lippincott. New York: Pocket Books, 224.

Wood, B. M., J. D. Negrey, J. L. Brown, T. Deschner, M. E. Thompson, S. Gunter, J. C. Mitani, D. P. Watts, and K. E. Langergraber. 2023. "Demographic and hormonal evidence for menopause in wild chimpanzees." *Science* 382 (6669): eadd5473. https://doi.org/10.1126/science.add5473.

Wood, J. W. 1989. "Fecundity and natural fertility in humans." *Oxf Rev Reprod Biol* 11: 61–109.

Wood, J. W. 1994. "Ch. 9 Menarche and menopause." In *Dynamics of Human Reproduction: Biology, Biometry, Demography*, Aldine De Gruyter (ed.), 401–439. New York: Aldine De Gruyter.

World Bank Group. 2024. "Current health expenditure per capita." *World Bank*. Last Modified 15 Apr 2024. Accessed 27 May 2025. https://data.worldbank.org/indicator/SH.XPD.CHEX.PC.CD.

Wrangham, R. W. 2019. "Hypotheses for the evolution of reduced reactive aggression in the context of human self-domestication." *Front Psychol* 10: 1914. https://doi.org/10.3389/fpsyg.2019.01914.

Wrotek, S., E. K. LeGrand, A. Dzialuk, and J. Alcock. 2021. "Let fever do its job: the meaning of fever in the pandemic era." *Evol Med Public Health* 9 (1): 26–35. https://doi.org/10.1093/emph/eoaa044.

Yamada, K., N. Sato-Mito, J. Nagata, and K. Umegaki. 2008. "Health claim evidence requirements in Japan." *J Nutr* 138 (6): 1192s–1198s. https://doi.org/10.1093/jn/138.6.1192S.

Yamaguchi, M.. 2009. "Aso's kanji conundrums spur self-reflection, textbook sales." *The Japan Times*, 7 Mar. https://www.japantimes.co.jp/news/2009/03/07/national/asos-kanji-conundrums-spur-self-reflection-textbook-sales/.

Yao, R., C. He, and P. Xiao. 2023. "'Food and medicine continuum' in the East and West: old tradition and current regulation." *Chin Herb Med* 15 (1): 6–14. https://doi.org/10.1016/j.chmed.2022.12.002.

Yasui, T., Y. Ideno, H. Shinozaki, Y. Kitahara, K. Nagai, and K. Hayashi. 2022. "Prevalence of the use of oral contraceptives and hormone replacement therapy in Japan: the Japan nurses' health study." *J Epidemiol* 32 (3): 117–124. https://doi.org/10.2188/jea.JE20200207.

Yates-Doerr, E. 2017. "Where is the local?: partial biologies, ethnographic sitings." *HAU J Ethnogr Theory* 7 (2): 377–401. https://doi.org/10.14318/hau7.2.032.

Yoshino, T., K. Katayama, K. Munakata, Y. Horiba, R. Yamaguchi, S. Imoto, S. Miyano, and K. Watanabe. 2013. "Statistical analysis of hie (cold sensation) and hiesho (cold disorder) in kampo clinic." *Evid Based Complement Alternat Med* 2013: 398458. https://doi.org/10.1155/2013/398458.

Yudell, M., D. Roberts, R. DeSalle, and S. Tishkoff. 2016. "Taking race out of human genetics." *Science* 351 (6273): 564–565. https://doi.org/10.1126/science.aac4951.

Zahavi, A. 1975. "Mate selection—a selection for a handicap." *J Theor Biol* 53 (1): 205–214. https://doi.org/10.1016/0022-5193(75)90111-3.

Zerjal, T., Y. Xue, G. Bertorelle, R. Spencer Wells, W. Bao, S. Zhu, R. Qamar, Q. Ayub, A. Mohyuddin, S. Fu, P. Li, N. Yuldasheva, R. Ruzibakiev, J. Xu, Q. Shu, R. Du, H. Yang, M. E. Hurles, E. Robinson, T. Gerelsaikhan, B. Dashnyam, S. Qasim Mehdi, and C. Tyler-Smith. 2003. "The genetic legacy of the Mongols." *Am J Hum Genet* 72 (3): 717–721. https://doi.org/10.1086/367774.

Zeserson, J. M. 2001. "How Japanese women talk about hot flushes: implications for menopause research." *Med Anthropol Q* 15 (2): 189–205. https://doi.org/10.1525/maq.2001.15.2.189.

Zhang, W. B., K. Ye, N. Barzilai, and S. Milman. 2021. "The antagonistic pleiotropy of insulin-like growth factor 1." *Aging Cell* 20 (9): e13443. https://doi.org/10.1111/acel.13443.

Zhou, Y. 2015. "Career interruption of Japanese women: why is it so hard to balance work and childcare?" *Japan Lab Rev* 12 (2): 106–123. https://www.jil.go.jp/english/JLR/documents/2015/JLR46_zhou.pdf.

Zipple, M. N., J. Altmann, F. A. Campos, M. Cords, L. M. Fedigan, R. R. Lawler, E. V. Lonsdorf, S. Perry, A. E. Pusey, T. S. Stoinski, K. B. Strier, and S. C. Alberts. 2021. "Maternal death and offspring fitness in multiple wild primates." *Proc Natl Acad Sci* 118 (1): e2015317118. https://doi.org/10.1073/pnas.2015317118.

INDEX

Note: **Bold** page numbers refer to tables and *italic* page numbers refer to figures.

For Product Safety Concerns and Information please contact our EU representative GPSR@taylorandfrancis.com
Taylor & Francis Verlag GmbH, Kaufingerstraße 24, 80331 München, Germany

www.ingramcontent.com/pod-product-compliance
Ingram Content Group UK Ltd.
Pitfield, Milton Keynes, MK11 3LW, UK
UKHW022317250926
475702UK00046B/901
9781032328584